HOUGHTON MIFFLIN HARCOURT

Go Math!

Intensive Intervention

RtI Response to Intervention Tier 3

Skill Packs

Grade 6

Houghton
Mifflin
Harcourt

www.hmhschool.com

Copyright © by Houghton Mifflin Harcourt Publishing Company.

All rights reserved. No part of this work may be reproduced or transmitted in any form or by any means, electronic or mechanical, including photocopying or recording, or by any information storage or retrieval system, without the prior written permission of the copyright owner unless such copying is expressly permitted by federal copyright law.

Permission is hereby granted to individuals using the corresponding student's textbook or kit as the major vehicle for regular classroom instruction to photocopy copying masters from this publication in classroom quantities for instructional use and not for resale. Requests for information on other matters regarding duplication of this work should be addressed to Houghton Mifflin Harcourt Publishing Company, Attn: Contracts, Copyrights, and Licensing, 9400 Southpark Center Loop, Orlando, Florida 32819-8647.

Printed in the U.S.A.

ISBN 978-0-544-24919-6

3 4 5 6 7 8 9 10   0928   22 21 20 19 18 17 16 15 14 13

4500452027   A B C D E F G

If you have received these materials as examination copies free of charge, Houghton Mifflin Harcourt Publishing Company retains title to the materials and they may not be resold. Resale of examination copies is strictly prohibited.

Possession of this publication in print format does not entitle users to convert this publication, or any portion of it, into electronic format.

Contents

Skills

1	Place Value Through Hundred Thousands	IIN1
2	Compare and Order Whole Numbers	IIN3
3	Round Whole Numbers	IIN5
4	Add and Subtract 4-Digit Numbers	IIN7
5	Multiplication and Division Facts	IIN9
6	Practice the Facts	IIN11
7	Estimate Products	IIN13
8	Multiply by 1- and 2-Digit Numbers	IIN15
9	Multiply Money	IIN17
10	Estimate Quotients	IIN19
11	Place the First Digit	IIN21
12	Factors	IIN23
13	Common Factors	IIN25
14	Multiples	IIN27
15	Understand Fractions	IIN29
16	Compare Fractions	IIN31
17	Compare and Order Fractions	IIN33
18	Model Equivalent Fractions	IIN35
19	Understand Mixed Numbers	IIN37
20	Multiply or Divide to Find Equivalent Fractions	IIN39
21	Simplest Form	IIN41
22	Add and Subtract Fractions	IIN43
23	Rename Fractions and Mixed Numbers	IIN45
24	Relate Fractions and Decimals	IIN47
25	Decimal Models	IIN49
26	Compare Decimals	IIN51
27	Multiply Decimals by Whole Numbers	IIN53
28	Understand Percent	IIN55

29	Read a Frequency Table	IIN57
30	Mean	IIN59
31	Read a Pictograph	IIN61
32	Read a Bar Graph	IIN63
33	Circle Graphs	IIN65
34	Addition Properties	IIN67
35	Multiplication Properties	IIN69
36	Expressions	IIN71
37	Number Patterns	IIN73
38	Patterns and Functions	IIN75
39	Geometric Patterns	IIN77
40	Use a Coordinate Grid	IIN79
41	Points, Lines, and Rays	IIN81
42	Angles	IIN83
43	Classify Angles	IIN85
44	Polygons	IIN87
45	Triangles	IIN89
46	Quadrilaterals	IIN91
47	Circles	IIN93
48	Congruent and Similar Figures	IIN95
49	Symmetry	IIN97
50	Transformations	IIN99
51	Faces of Solid Figures	IIN101
52	Faces, Edges, and Vertices	IIN103
53	Choose the Appropriate Unit	IIN105
54	Customary Units of Capacity	IIN107
55	Metric Units of Capacity	IIN109
56	Read a Thermometer	IIN111
57	Perimeter	IIN113
58	Estimate and Find Area	IIN115

59	Multiply with Three Factors	IIN117
60	Explore Volume	IIN119
61	More Likely, Less Likely, Equally Likely	IIN121
62	Tree Diagrams	IIN123

Name_____

Place Value Through Hundred Thousands
Skill 1

Learn the Math

You can use base-ten blocks to model, read, and write whole numbers.

There are 2,737 seats in the New York State Theater. Use base-ten blocks to show the number of seats.

Vocabulary
- digit
- place value
- expanded form
- standard form
- word form

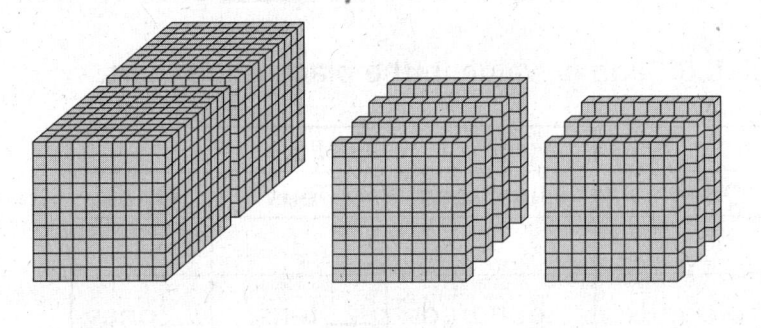

__2__ thousands _____ hundreds _____ tens _____ ones

Write: 2,737

Read: two thousand, seven hundred thirty-seven

What is the value of each digit in the number 324,856?

The value of a digit depends on its place-value position in the number.

You can use a place-value chart to show the value of each digit in a number.

THOUSANDS			ONES		
Hundreds	Tens	Ones	Hundreds	Tens	Ones
3	2	4	8	5	6
3 hundred thousands	2 ten thousands	4 thousands	8 hundreds	5 tens	6 ones
3 × 100,000	2 × 10,000	4 × 1,000	8 × 100	5 × 10	6 × 1
300,000	20,000	4,000	800	50	6

Standard form: 324,856

Word form: three hundred twenty-four thousand, eight _____ fifty-_____

Expanded form: 300,000 + 20,000 + _____ + 800 + _____ + 6

Response to Intervention • Tier 3   IIN1

Do the Math

Skill 1

Complete.

1. Write how many thousands, hundreds, tens, and ones the model shows.

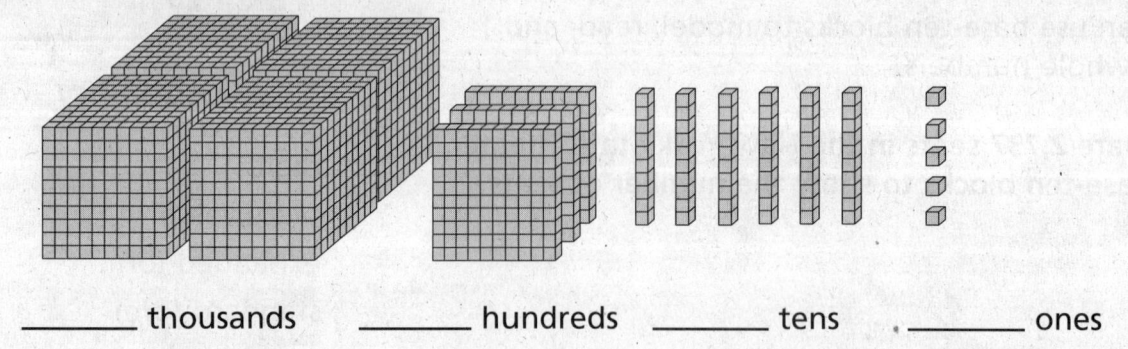

   _____ thousands _____ hundreds _____ tens _____ ones

2. Use the number 95,703. Write each digit and its value in the place-value chart.

| THOUSANDS ||| ONES |||
Hundreds	Tens	Ones	Hundreds	Tens	Ones
	___ten thousands	___thousands	___hundreds	___ tens	___ones

Check

3. The model shows the number of students at an elementary school. Explain why the model does NOT represent 3,140.

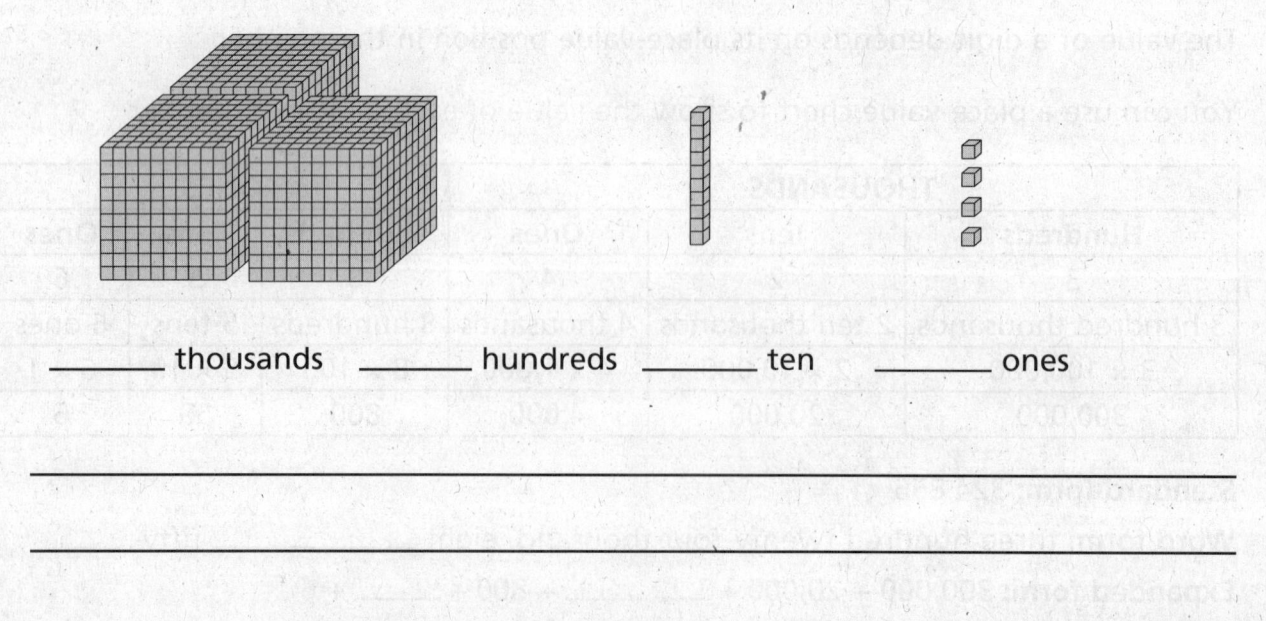

   _____ thousands _____ hundreds _____ ten _____ ones

   _____
   _____
   _____

IIN2 Response to Intervention • Tier 3

Name_____

Compare and Order Whole Numbers
Skill 2

Learn the Math

You can compare and order numbers in different ways.

Use base-ten blocks.
Compare 1,343 and 1,234. Write <, >, or =.

Vocabulary
equal to (=)
greater than (>)
less than (<)
place value

thousands   hundreds   tens   ones

Model 1,343.

Model 1,234.

1 = 1   3 > 2

So, 1,343 ◯ 1,234.

Use a place-value chart. Compare 4,259,178 and 4,284,307. Write <, >, or =.

Start at the left and compare the digits in each place-value position until the digits differ.

MILLIONS			THOUSANDS			ONES		
Hundreds	Tens	Ones	Hundreds	Tens	Ones	Hundreds	Tens	Ones
		4,	2	5	9,	1	7	8
		4,	2	8	4,	3	0	7

4◯4   2◯2   5◯8

So, 4,259,178 ◯ 4,284,307.

Use a number line. Order 23,427; 23,297; 23,320 from least to greatest.
The farther to the right on the number line, the greater the number.

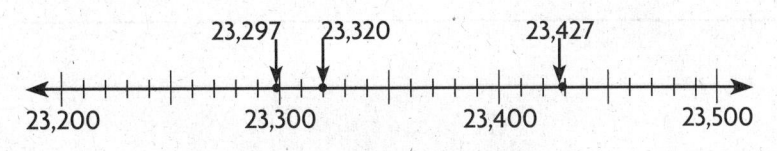

23,297   23,320   23,427

23,200   23,300   23,400   23,500

23,297 is to the left of 23,320.
23,320 is to the left of 23,427.

So, 23,297 ◯ 23,320 ◯ 23,427.

Response to Intervention • Tier 3   IIN3

Do the Math　　　　　　　　　　　　　　　　　　　　　　**Skill ❷**

Compare. Write <, >, or = for each ◯.

1. 295 ◯ 295

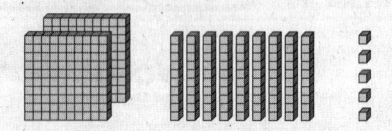

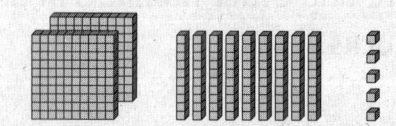

2. 415,878 ◯ 451,686

THOUSANDS			ONES		
Hundreds	Tens	Ones	Hundreds	Tens	Ones
_____	_____	_____	_____	_____	_____
_____	_____	_____	_____	_____	_____

3. 6,527,117 ◯ 6,517,710

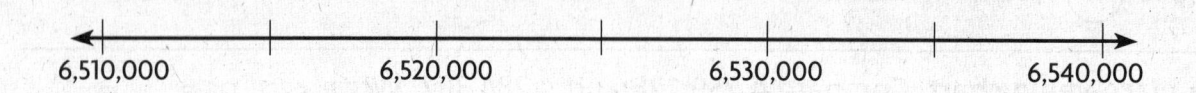

|←————|————|————|————|————→|
6,510,000　6,520,000　　6,530,000　　6,540,000

Use the number line below to write the numbers in order from greatest to least.

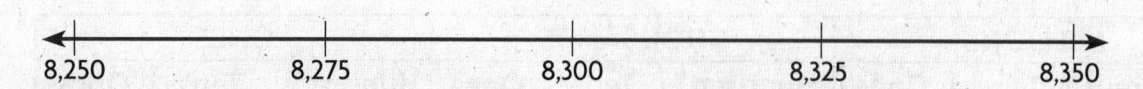

|←————|————|————|————|————→|
8,250　　8,275　　8,300　　8,325　　8,350

4. 8,325; 8,290; 8,349 _____

5. 8,289; 8,302; 8,283 _____

Check

6. Explain why you do NOT start with the first place-value position on the right when comparing numbers.

_____
_____
_____
_____

IIN4　Response to Intervention • Tier 3

Name_____

Round Whole Numbers
Skill 3

Learn the Math

You can round numbers in different ways.

Use a number line.
Round 17,685 to the nearest hundred.

Step 1 Find 17,685 on the number line.

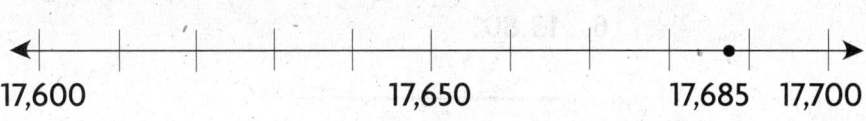

```
←——|——|——|——|——|——|——|——•—|——→
17,600        17,650           17,685 17,700
```

Vocabulary
equal to (=)
greater than (>)
less than (<)
number line
round

Step 2 Decide which hundred 17,685 is closer to.
17,685 is closer to 17,700 than to 17,600.

So, 17,685 rounds to 17,700.

Use rounding rules.
Round 4,314,213 to the nearest thousand.

Step 1 Find the place to which you want to round.

    4,314,213   **Think:** 4 is in the thousands place.
                             Round to 4,314,000 or 4,315,000.

Step 2 Look at the digit to the right.

    4,314,<u>2</u>13   Is 2 less than, greater than, or equal to 5?
                      If the digit is less than 5, the digit in the rounding place
                      stays the same. If the digit is greater than or equal to 5,
                      the digit in the rounding place increases by one.

Since 2 < 5, the digit 4 stays the same.

Step 3 Change each digit to the right of the rounding place to zero.

So, 4,314,213 rounds to _____.

Response to Intervention • Tier 3   IIN5

Do the Math

Skill 3

Round each number to the place value of the underlined digit.

1. 9<u>2</u>1
   _____

2. <u>4</u>69
   _____

3. 2,8<u>7</u>6
   _____

4. 5,<u>1</u>99
   _____

5. <u>3</u>,321
   _____

6. 13,8<u>0</u>1
   _____

7. 18,<u>7</u>11
   _____

8. 1<u>6</u>,847
   _____

9. 215,2<u>9</u>3
   _____

10. 494,<u>5</u>34
    _____

11. 9<u>9</u>2,429
    _____

12. 1,813,4<u>2</u>9
    _____

13. 6,719,<u>2</u>10
    _____

14. 5,2<u>0</u>3,112
    _____

Check

15. Explain how to round 1,670 to the nearest thousand.

_____

_____

_____

_____

Add and Subtract 4-Digit Numbers
Skill 4

Learn the Math

Add. 3,847 + 1,258

Step 1	Step 2	Step 3	Step 4
Add the ones.	Add the tens.	Add the hundreds.	Add the thousands.
Regroup 15 ones as 1 ten 5 ones.	Regroup 10 tens as 1 hundred 0 tens.	Regroup 11 hundreds as 1 thousand 1 hundred.	
1 $3,847$ $+1,258$ $\overline{5}$	11 $3,847$ $+1,258$ $\overline{05}$	$1\,1\,1$ $3,847$ $+1,258$ $\overline{105}$	$1\,1\,1$ $3,847$ $+1,258$ $\overline{5,105}$

So, 3,847 + 1,258 = _____.

Subtract. 6,138 − 3,249

Step 1	Step 2	Step 3	Step 4
Subtract the ones.	Subtract the tens.	Subtract the hundreds.	Subtract the thousands.
Since 9 > 8, regroup 3 tens 8 ones as 2 tens 18 ones.	Since 4 > 2, regroup 1 hundred 2 tens as 0 hundreds 12 tens.	Since 2 > 0, regroup 6 thousands 0 hundreds as 5 thousands 10 hundreds.	
$2\,18$ $6,1\cancel{3}\cancel{8}$ $-3,249$ $\overline{9}$	12 $0\,\cancel{}\,18$ $6,\cancel{1}\cancel{3}\cancel{8}$ $-3,249$ $\overline{89}$	$10\,12$ $5\,\,\cancel{0}\,\cancel{}\,18$ $\cancel{6},\cancel{1}\cancel{3}\cancel{8}$ $-3,249$ $\overline{889}$	$10\,12$ $5\,\,\cancel{0}\,\cancel{}\,18$ $\cancel{6},\cancel{1}\cancel{3}\cancel{8}$ $-3,249$ $\overline{2,889}$

So, 6,138 − 3,249 = _____.

Response to Intervention • Tier 3   IIN7

Do the Math   **Skill 4**

Find the sum or difference.

1. 5,424
   + 1,597
   _____

2. 1,681
   + 1,299
   _____

3. 3,881
   + 2,325
   _____

4. 1,294
   + 6,468
   _____

5. 6,093
   − 2,470
   _____

6. 1,411
   + 1,027
   _____

7. 8,629
   − 3,859
   _____

8. 5,726
   + 3,618
   _____

Check

9. Look at the problem below. What is the missing number if you do NOT need to regroup to subtract? Explain.

    7,35■
   − 4,148
   _____

_____

_____

IIN8   Response to Intervention • Tier 3

Name_____

Multiplication and Division Facts
Skill 5

Learn the Math

Strategies can help you learn multiplication and division facts.

Use the break apart strategy.
Multiply. 7 × 8

Vocabulary
array

Step 1	Step 2	Step 3
Draw an array that is 7 units wide and 8 units long. The area of the array is 7 × 8.	Break apart the array to make two smaller arrays for products you know. **Think:** 8 = 4 + 4	Find the sum of the products of the two smaller arrays.
8 wide, 7 tall array	4 and 4 arrays, 7 tall	7 × 4 = _____ 7 × 4 = _____ _____ + _____ = _____

So, 7 × 8 = _____ .

Use skip-counting.
Multiply. 9 × 5

Skip-count by 5 until you reach the ninth value.

5,   10,   15,   20,   25,   30,   35,   40,   45

So, 9 × 5 = _____ .

Use inverse operations.
Divide. 48 ÷ 8

Use a multiplication fact to find a division fact.

Think: What number times 8 equals 48? _____

_____ × 8 = 48

So, 48 ÷ 8 = _____ .

Response to Intervention • Tier 3   IIN9

Do the Math

Skill 5

Find the product. Show the strategy you used.

1. 7 × 7 = _____
2. 8 × 5 = _____

3. 6 × 9 = _____
4. 10 × 3 = _____

5. 7 × 5 = _____
6. 11 × 4 = _____

Find the quotient. Show the strategy you used.

7. 42 ÷ 6 = _____
8. 81 ÷ 9 = _____

9. 7)63
10. 5)50

11. 8)24
12. 9)36

Check

13. What strategy did you use to solve Problem 8?
    Why did you choose that strategy?

    _____

    _____

    _____

IIN10   Response to Intervention • Tier 3

Name_____

Practice the Facts
Skill 6

Learn the Math

You can use a number line to practice basic facts.

Use a number line to add. 3 + 9 = __?__

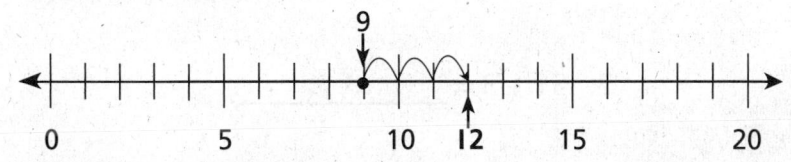

Vocabulary

number line

Start with the greater number. Find 9 on the number line.
Count 3 places to the right.

So, 3 + 9 = _____ .

Use a number line to subtract. 16 − 8 = __?__

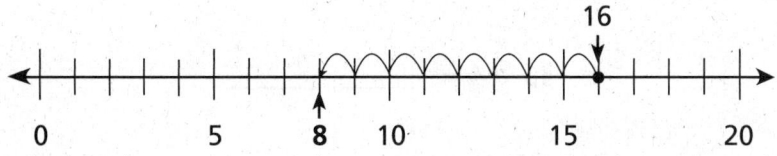

Find 16 on the number line. Count 8 places to the left.

So, 16 − 8 = _____ .

Use a number line to multiply. 5 × 6 = __?__

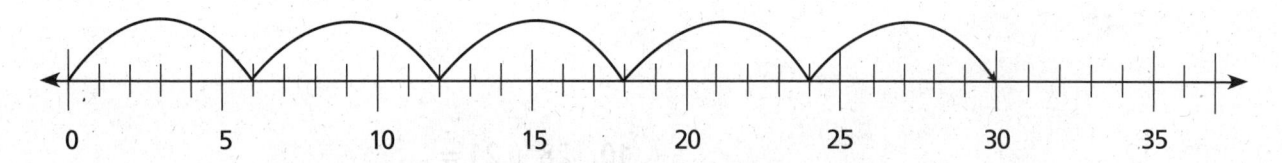

Start at 0. Make 5 jumps of 6. Which number do you land on? _____

So, 5 × 6 = _____ .

Use a number line to divide. 27 ÷ 9 = __?__

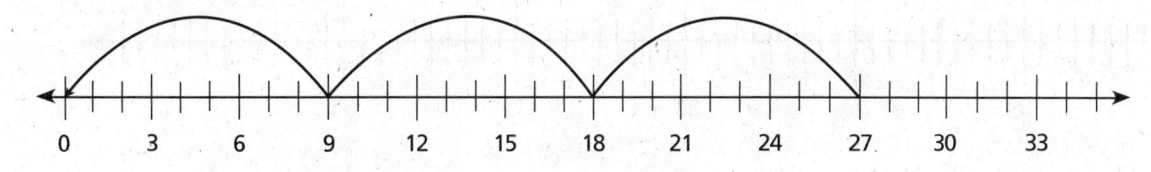

Find 27 on the number line. Jump back 9 spaces at a time, until you reach 0.
How many jumps did you make? _____

So, 27 ÷ 9 = _____ .

Response to Intervention • Tier 3   IIN11

Do the Math

Skill 6

Use a number line to find the sum, difference, product, or quotient.
Use the number line below or draw your own.

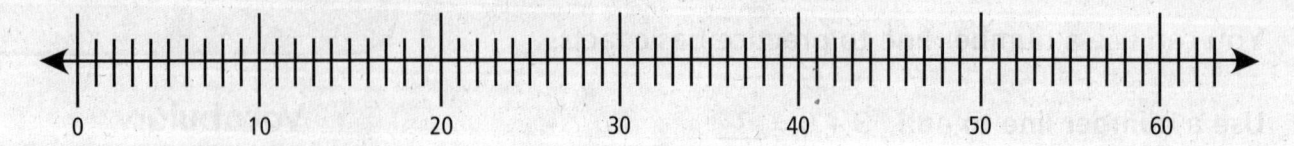

[number line from 0 to 60]

1. 7 × 6 = _____

2. 51 ÷ 3 = _____

3. 12 + 15 = _____

4. 40 − 9 = _____

5. 8 × 5 = _____

6. 36 ÷ 4 = _____

7. 11 − 3 = _____

8. 25 + 25 = _____

9. 40 − 12 = _____

10. 28 + 21 = _____

Check

11. Explain how you would use the number line to divide 49 ÷ 7.

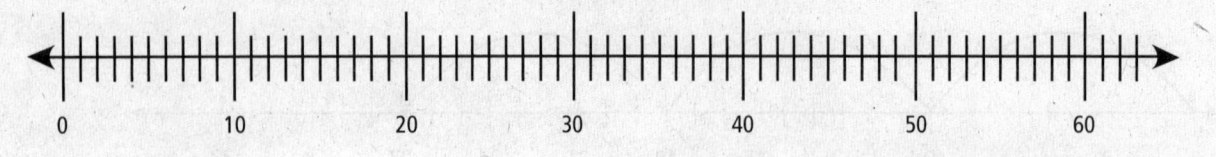

[number line from 0 to 60]

_____
_____
_____

IIN12   Response to Intervention • Tier 3

Name_____

Estimate Products
Skill 7

Learn the Math

You can estimate products by rounding or by using compatible numbers.

Vocabulary

compatible numbers
estimate
product

Use rounding and mental math.
Estimate. 43 × 387

Step 1   Round each factor to the nearest place value.

43 rounded to the nearest ten is 40.

387 rounded to the nearest hundred is 400.

43 × 387 ⟶ 40 × 400

Step 2   Use mental math.
Think of a basic fact you can use to solve.
Then use a pattern.

$4 \times 4 = 16$
$40 \times 40 = 1{,}600$
$40 \times 400 = $ _____

So, 43 × 387 is about _____ .

Use compatible numbers and mental math.
Estimate. 29 × 321

Step 1   Look at both factors.
Use multiplication facts you know.
Think: 30 × 300 should be easier to compute mentally.

Step 2   Use mental math.

$3 \times 30 = $ _____
$30 \times 30 = $ _____
$30 \times 300 = $ _____

So, 29 × 321 is about _____ .

Response to Intervention • Tier 3   IIN13

Do the Math Skill 7

Use rounding to estimate the product. Show the factors you used to estimate.

1. 19 × 52

2. 37 × 64

3. 78 × 410

4. 23 × 678

5. 49 × 93

6. 55 × 625

Use compatible numbers to estimate the product. Show the compatible numbers you used to estimate.

7. 22 × 249

8. 34 × 186

9. 17 × 375

10. 29 × 752

11. 39 × 147

12. 12 × 119

Check

13. Explain one way to estimate the product of 36 × 195.

_____
_____
_____
_____

IIN14   Response to Intervention • Tier 3

Name _____

Multiply by 1- and 2-Digit Numbers

Skill 8

Learn the Math

You can use different strategies to multiply numbers.

Use a model. Multiply. 6 × 17

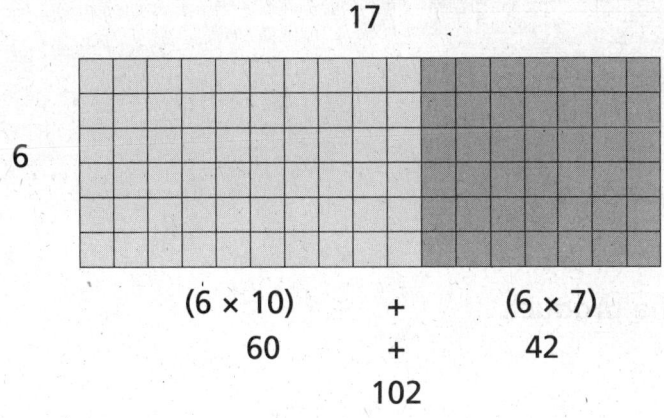

Vocabulary

factor
partial products
regroup

17

6

(6 × 10) + (6 × 7)
   60   +   42
          102

So, 6 × 17 = 102.

Use partial products. Multiply. 23 × 59

Step 1 Multiply the ones.
3 × 9 = 27   3 × 50 = 150

Step 2 Multiply the tens.
20 × 9 = 180   20 × 50 = 1,000

Step 3 Add the partial products.
27 + 150 + 180 + 1,000 = 1,357

So, 23 × 59 = _____ .

```
    59
 ×  23
 ─────
    27  ← 3 × 9
   150  ← 3 × 50
   180  ← 20 × 9
 +1,000 ← 20 × 50
 ─────
```

Use regrouping. Multiply. 5 × 248

Step 1 Multiply the ones.   5 × 8 ones = 40 ones
Regroup 40 ones as 4 tens 0 ones.

Step 2 Multiply the tens.   5 × 4 tens = 20 tens
Add the regrouped tens.   20 tens + 4 tens = 24 tens
Regroup 24 tens as 2 hundreds 4 tens.

Step 3 Multiply the hundreds.   5 × 2 hundreds = 10 hundreds
Add the regrouped hundreds.
10 hundreds + 2 hundreds = 12 hundreds
Regroup 12 hundreds as 1 thousand 2 hundreds.

```
   2 4
   248
 ×   5
 ─────
 1,240
```

So, 5 × 248 = _____ .

Response to Intervention • Tier 3   IIN15

Do the Math

Skill 8

Multiply. Use the model to find the product.

1. 4 × 13

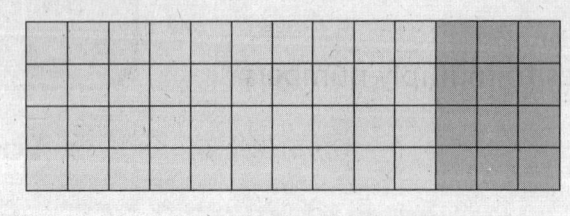

(___ × ___) + (___ × ___)

___ + ___

___

Multiply. Use partial products to find the product.

2. 23
   × 22

3. 35
   × 18

4. 41
   × 7

Multiply. Use regrouping to find the product.

5. 72
   × 8

6. 26
   × 5

7. 238
   × 14

Check

8. Look at the problem below. How do you record the 28 ones when you use regrouping to find the product of 4 ones and 7 ones?

    47
   × 14

_____

_____

_____

IIN16   Response to Intervention • Tier 3

Name_____

Multiply Money
Skill 9

Learn the Math

Multiply money the same way you multiply whole numbers. Align the digits by place value. Regroup as needed.

Vocabulary
decimal point
partial product
place value
regroup

Multiply. $1.85 × 24

Step 1 Multiply by the ones.

```
       3 2
     $1.85
   ×    24
      740    ← 4 × 185
```

Step 2 Multiply by the tens.

```
       1 1
       3̶ 2̶
     $1.85
   ×    24
       740
      3700   ← 20 × 185
```

Step 3 Add the partial products.

```
     $1.85
   ×    24
       740
   + 3700
      4440
```

Step 4 Place the decimal point two places to the left to show dollars and cents.

```
     $1.85
   ×    24
       740
   + 3700
    $44.40
```

So, $1.85 × 24 = _____.

Response to Intervention • Tier 3   IIN17

Do the Math

Skill 9

Find the product. Show your work.

1. $1.55
    × 18

2. $4.37
    × 46

3. $5.73
    × 26

4. $2.28
    × 49

5. $6.40
    × 25

6. $3.99
    × 52

7. $4.62
    × 35

8. $2.84
    × 70

9. $7.49
    × 22

10. $9.95
    × 17

Check

11. Explain how you know where to put the decimal point in a multiplication problem with money. What other symbol do you need to use? Why?

_____
_____
_____

IIN18 Response to Intervention • Tier 3

Name_____

Estimate Quotients
Skill ⑩

Learn the Math

You can estimate quotients by rounding or by using compatible numbers.

Vocabulary
compatible numbers
dividend
estimate
round

Use rounding and mental math.
Estimate. 8,792 ÷ 3

Step 1 Round the dividend to the nearest thousand.	8,792 ÷ 3 → 9,000 ÷ 3
Step 2 Use mental math. Think of a basic fact you can use to solve.	9 ÷ 3 = 3
Step 3 Use a pattern to estimate.	90 ÷ 3 = 30 900 ÷ 3 = 300 9,000 ÷ 3 = _____

So, 8,792 ÷ 3 is about _____.

Use compatible numbers and mental math.
Estimate. 2,630 ÷ 5

Step 1 Look at the first two digits of the dividend: **2,630**. Use division facts for 5 to find a basic fact that is close to **26 ÷ 5**.	Think: 25 ÷ 5 = 5 and 30 ÷ 5 = 6 So, you can use 2,500 or 3,000 for 2,630.
Step 2 Use mental math.	2,500 ÷ 5 = _____ 3,000 ÷ 5 = _____

So, both _____ and _____ are reasonable estimates.

Response to Intervention • Tier 3   IIN19

Do the Math

Skill 10

Estimate the quotient. Use rounding and mental math.

1. 4,115 ÷ 8	2. 1,945 ÷ 2	3. 986 ÷ 5

   _____	 _____	 _____

4. 5,881 ÷ 3	5. 2,186 ÷ 2	6. 6,881 ÷ 7

   _____	 _____	 _____

Estimate the quotient. Use compatible numbers and mental math.

7. 245 ÷ 6	8. 2,139 ÷ 3	9. 1,742 ÷ 6

   _____	 _____	 _____

10. 637 ÷ 8	11. 4,312 ÷ 6	12. 297 ÷ 7

   _____	 _____	 _____

Check

13. Why is rounding to the nearest thousand NOT a good method for estimating the quotient for 1,396 ÷ 7?

_____

_____

_____

IIN20 Response to Intervention • Tier 3

Name_____

Place the First Digit
Skill 11

Learn the Math

You can use an estimate or place value to place the first digit in the quotient.

Use compatible numbers to estimate to place the first digit in the quotient.

Divide 172 by 4. Write $4\overline{)172}$.

Step 1	Step 2	Step 3
Use compatible numbers to estimate. Think: $4\overline{)160}^{\ 40}$ or $4\overline{)200}^{\ 50}$ $4\overline{)172}^{\ \blacksquare}$ So, the first digit is in the tens place.	Divide the 17 tens. $4\overline{)172}^{\ 4}$ Divide. $4\overline{)17}$ $\underline{-16}$ Multiply. $4 \times 4 = 16$ $\ \ \ 1$ Subtract. $17 - 16 = 1$ Compare. $1 < 4$	Bring down the 2 ones. Divide the 12 ones. $4\overline{)172}^{\ 43}$ Divide. $4\overline{)12}$ $\underline{-16}\!\downarrow$ Multiply. $4 \times 3 = 12$ $\ \ 12$ Subtract. $12 - 12 = 0$ $\underline{-12}$ Compare. $0 < 4$ $\ \ \ 0$

So, 172 divided by 4 is _____.

Use place value to place the first digit in the quotient.

Divide 380 by 5. Write $5\overline{)380}$.

Step 1	Step 2	Step 3
Look at the hundreds. $5\overline{)380}$ $3 < 5$ You cannot divide 5 into 3, so look at the tens. $5\overline{)380}^{\ \blacksquare}$ $38 > 5$ Place the first digit in the tens place.	Divide the 38 tens. $5\overline{)380}^{\ 7}$ Divide. $5\overline{)38}$ $\underline{-35}$ Multiply. $5 \times 7 = 35$ $\ \ \ 3$ Subtract. $38 - 35 = 3$ Compare. $3 < 5$	Bring down the 0 ones. Divide the 30 ones. $5\overline{)380}^{\ 76}$ Divide. $5\overline{)30}$ $\underline{-35}\!\downarrow$ Multiply. $5 \times 6 = 30$ $\ \ 30$ Subtract. $30 - 30 = 0$ $\underline{-30}$ Compare. $0 < 5$ $\ \ \ 0$

So, 380 divided by 5 is _____.

Do the Math

Skill 11

Tell where to place the first digit. Then divide.

1. 2)438 _____ place

2. 9)918 _____ place

3. 5)105 _____ place

4. 7)574 _____ place

5. 4)956 _____ place

6. 9)315 _____ place

7. 5)825 _____ place

8. 8)896 _____ place

Check

9. Look at Problem 7. Where would you place the first digit if the dividend were 225 instead of 825? Explain.

_____

_____

_____

IIN22   Response to Intervention • Tier 3

Name_____

Factors
Skill ⑫

Learn the Math

A factor is a number multiplied by another number to find a product. Every whole number has at least two factors, 1 and itself.

$$1 \times 11 = 11$$
factor factor product

Vocabulary
array
factor

Many numbers can be broken into factors in different ways.

9 = 1 × 9   9 = 9 × 1   9 = 3 × 3

Make all the arrays you can with 10 square tiles to show all the factors of 10.

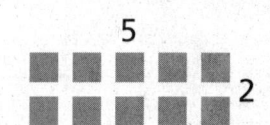

10 rows of 1   1 row of 10   5 rows of 2   2 rows of 5
10 × 1 = 10   1 × 10 = 10   5 × 2 = 10   2 × 5 = ____
factors = ____, ____   factors = 1, 10   factors = ____, ____   factors = ____, ____

Write the factors in order from least to greatest.
Each factor should be written once.
So, the factors of 10 are _____.

Response to Intervention • Tier 3   IIN23

Do the Math

Skill 12

Use arrays to find all of the factors of each product.

1. 12

    Factors = _____

2. 30

    Factors = _____

3. 24

    Factors = _____

4. 18

    Factors = _____

5. 40

    Factors = _____

6. 10

    Factors = _____

7. 36

    Factors = _____

8. 21

    Factors = _____

9. 44

    Factors = _____

10. 41

    Factors = _____

Check

11. Explain how you can use arrays to find the factors of a number.

    _____
    _____
    _____
    _____
    _____

Name_____

Common Factors

Skill 13

Learn the Math

A factor is a number multiplied by another number to find a product. A number that is a factor of two or more numbers is a common factor.

Vocabulary

common factor

Find the common factors of 14 and 21.

Step 1 List all the factors of 14.

　　1 × 14 = 14　　　2 × 7 = 14

　　So, the factors of 14 are 1, 2, 7, and 14.

Step 2 List all of the factors of 21.

　　1 × 21 = 21　　　3 × 7 = 21

　　So, the factors of 21 are 1, 3, 7, and 21

Step 3 Circle the factors that appear in both lists.

　　Factors of 14: ①, 2, ⑦, 14

　　Factors of 21: ①, 3, ⑦, 21

So, the common factors of 14 and 21 are ____ and ____ .

Find the common factors of 10 and 18.

Step 1 List all the factors of 10.

　　1 × 10 = 10　　　_____

　　So, the factors of 10 are 1, ____, ____, and 10.

Step 2 List all of the factors of 18.

　　1 × 18 = 18　　　_____　　_____

　　So, the factors of 18 are 1, ____, ____, ____, ____, and 18.

Step 3 Circle the factors that appear in both lists.

　　Factors of 10: 1, 2, 5, 10

　　Factors of 18: ____, ____, ____, ____, ____, ____ .

　　The common factors of 10 and 18 are ____ and ____ .

Response to Intervention • Tier 3　IIN25

Do the Math Skill 13

List all the factors for each pair of numbers. Then identify the common factors.

1. 3, 17

   Factors of 3: ___, ___

   Factors of 17: ___, ___

   Common Factors: ___

2. 12, 16

   Factors of 12: ___, ___, ___, ___, ___, ___

   Factors of 16: ___, ___, ___, ___, ___

   Common Factors: ___, ___, ___

3. 15, 25

   Factors of 15: ___, ___, ___, ___

   Factors of 25: ___, ___, ___

   Common Factors: ___, ___

4. 10, 100

   Factors of 10: ___, ___, ___, ___

   Factors of 100: ___, ___, ___, ___, ___, ___, ___, ___, ___

   Common Factors: ___, ___, ___, ___

5. 28, 49

   Factors of 28: ___, ___, ___, ___, ___, ___

   Factors of 49: ___, ___, ___

   Common Factors: ___, ___

6. 11, 44

   Factors of 11: ___, ___

   Factors of 44: ___, ___, ___, ___, ___, ___

   Common Factors: ___, ___

7. 12, 20

   Factors of 12: ___, ___, ___, ___, ___, ___

   Factors of 20: ___, ___, ___, ___, ___, ___

   Common Factors: ___, ___, ___

8. 16, 27

   Factors of 16: ___, ___, ___, ___, ___

   Factors of 27: ___, ___, ___, ___

   Common Factors: ___

Check

9. Chris says the common factors for 14 and 28 are 1, 2, and 7. Are these all the common factors? If not, what is missing?

   _____

   _____

IIN26 Response to Intervention • Tier 3

Name_____

Multiples
Skill 14

Learn the Math

The product of two whole numbers is called a multiple of each of those numbers. The number of multiples a number has is endless.

You can make a model, skip-count, or multiply to find multiples.

Vocabulary

multiple
skip-count

Make a model.
List the first five multiples of 4.

Show 3 yellow counters and then 1 red counter. Repeat this pattern of 4 counters. Count the counters to find the multiples of 4. The red counters represent multiples of 4.

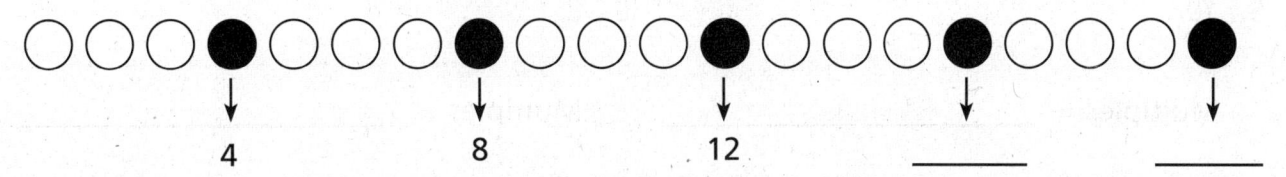

○○○● ○○○● ○○○● ○○○● ○○○●
↓   ↓   ↓   ↓   ↓
4   8   12   ____   ____

So, the first five multiples of 4 are _____.

Use a number line.
List the first five multiples of 5.

Skip-count by 5s on the number line. Name the multiples of 5.

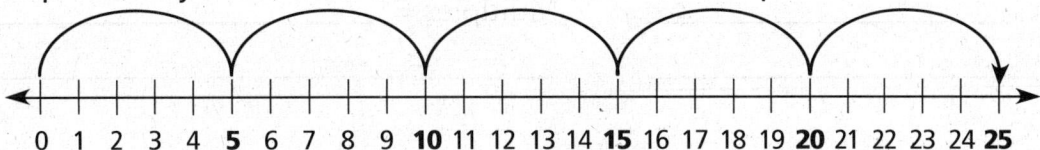

0 1 2 3 4 **5** 6 7 8 9 **10** 11 12 13 14 **15** 16 17 18 19 **20** 21 22 23 24 **25**

So, the first five multiples of 5 are _____.

Multiply.
List the first five multiples of 7.

Make a list. Multiply 7 by each of the first five whole numbers.

7 × 1	7 × 2	7 × 3	7 × 4	7 × 5
7	14			

So, the first five multiples of 7 are _____.

Response to Intervention • Tier 3   IIN27

Do the Math   Skill 14

List the first twelve multiples of each number.

1. 3

   Multiples = _____
   _____

2. 6

   Multiples = _____
   _____

3. 2

   Multiples = _____
   _____

4. 11

   Multiples = _____
   _____

5. 10

   Multiples = _____
   _____

6. 8

   Multiples = _____
   _____

7. 9

   Multiples = _____
   _____

8. 1

   Multiples = _____
   _____

Check

9. Tell whether 24 is a multiple of each number below. How do you know?
   1, 2, 3, 4, 5, 6, 7, 8, 9, 10, 11, 12

   _____
   _____
   _____
   _____

IIN28   Response to Intervention • Tier 3

Name_____

Understand Fractions
Skill 15

Learn the Math

A fraction is a number that names part of a whole or part of a group.

Write a fraction to name the shaded part.

number of shaded parts → $\underline{1}$ ← numerator
total equal parts → 3 ← denominator

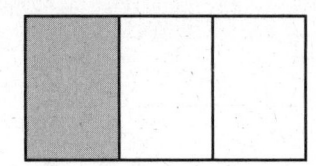

Read: one third
one out of three
one divided by three

Write: $\frac{1}{3}$

Vocabulary
denominator
fraction
numerator
unit fraction

Count equal parts of a whole.

Each equal part of the whole is $\frac{1}{6}$. The fraction $\frac{1}{6}$ is a unit fraction. It has a numerator of 1.

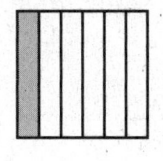

$\frac{1}{6}$ $\frac{}{6}$ $\frac{}{6}$ $\frac{}{6}$ $\frac{}{6}$ $\frac{}{6}$

$\frac{}{6}$ = one whole, or 1

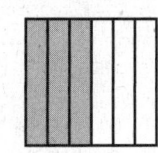

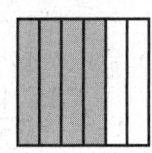

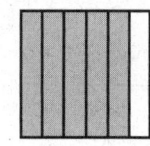

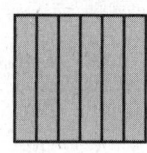

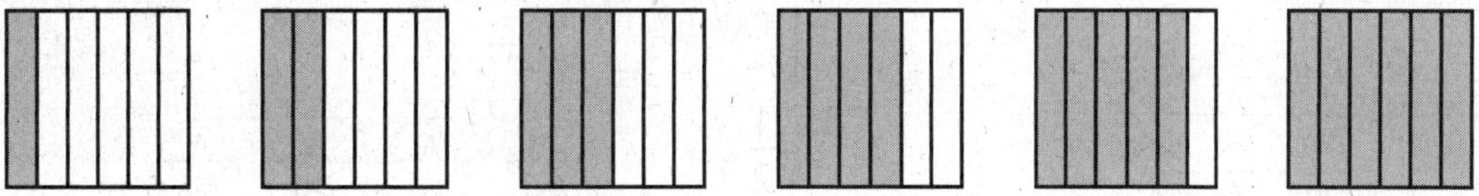

Write a fraction to name the shaded part of each group.

There are 4 equal parts.	There are 2 equal parts.	There are 3 equal parts.
1 out of 4 is shaded.	1 out of 2 is shaded.	2 out of 3 are shaded.
Read: one _____	**Read:** one _____	**Read:** two _____
Write: —	**Write:** —	**Write:** —

Response to Intervention • Tier 3   IIN29

Do the Math　　　　　　　　　　　　　　　　　　　　　　　　　　**Skill 15**

Write a fraction for the shaded part. Write a fraction for the unshaded part.

1. 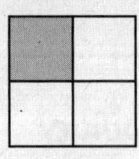 　　　　　　　　　　　　　　　　2.

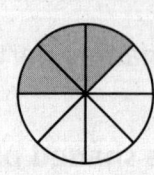

3.  　　　　　　　　　　　　　　　　4.

Shade the picture to show the fraction.

5. $\frac{2}{5}$　　　　　　　　　　　　　6. $\frac{4}{9}$

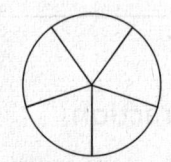

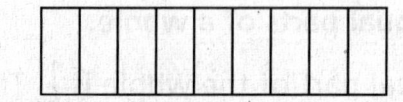

7. $\frac{3}{3}$ 　　　　　　　　　　　　　8. $\frac{1}{2}$

Write the fraction for each.

9. seven out of nine _____　　　10. four divided by five _____

11. three sixths _____　　　12. one eighth _____

Check

13. Zara drew some triangles on the board. She shaded 7 of the triangles, and left the last 4 triangles unshaded. She says that $\frac{4}{7}$ of the triangles are unshaded. Do you agree? Explain?

_____

_____

IIN30　Response to Intervention • Tier 3

Name_____

Compare Fractions
Skill 16

Learn the Math

You can use counters, fraction bars, and number lines to compare fractions.

Compare $\frac{1}{5}$ and $\frac{3}{5}$ using counters.

When comparing fractions with like denominators, only compare the numerators.

$\frac{1}{5}$ ●○○○○ $\frac{3}{5}$ ●●●○○

Compare the number of gray counters: 1 < 3, so $\frac{1}{5} < \frac{3}{5}$.

Vocabulary
denominator
greater than (>)
less than (<)
numerator

Compare $\frac{3}{4}$ and $\frac{5}{8}$ using fraction bars.

Line up three $\frac{1}{4}$ bars and five $\frac{1}{8}$ bars with the bar for 1.

1		
$\frac{1}{4}$	$\frac{1}{4}$	$\frac{1}{4}$
$\frac{1}{8}$ $\frac{1}{8}$ $\frac{1}{8}$ $\frac{1}{8}$ $\frac{1}{8}$		

Compare the two rows of fraction bars. The longer row represents the greater fraction.

So, $\frac{3}{4}$ ○ $\frac{5}{8}$, or $\frac{5}{8}$ ○ $\frac{3}{4}$.

Compare $\frac{7}{12}$ and $\frac{2}{3}$ using a number line.

Draw two number lines. Divide one into thirds and one into twelfths. Locate $\frac{2}{3}$ on one number line and $\frac{7}{12}$ on the other.

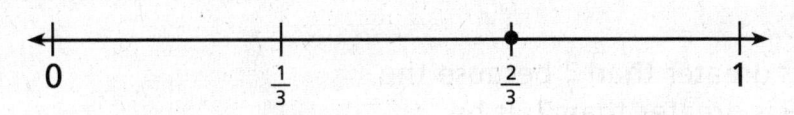

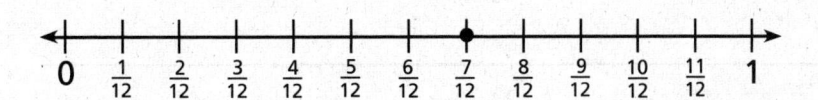

The fraction farther to the right is the greater fraction.

So, $\frac{7}{12}$ ○ $\frac{2}{3}$, or $\frac{2}{3}$ ○ $\frac{7}{12}$.

Do the Math

Skill 16

Compare. Write >, <, or =.

1. $\frac{6}{8} \bigcirc \frac{2}{8}$

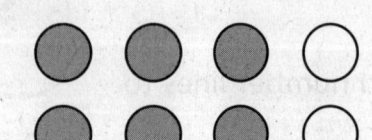

2. $\frac{4}{10} \bigcirc \frac{3}{5}$

3. $\frac{5}{6} \bigcirc \frac{6}{9}$

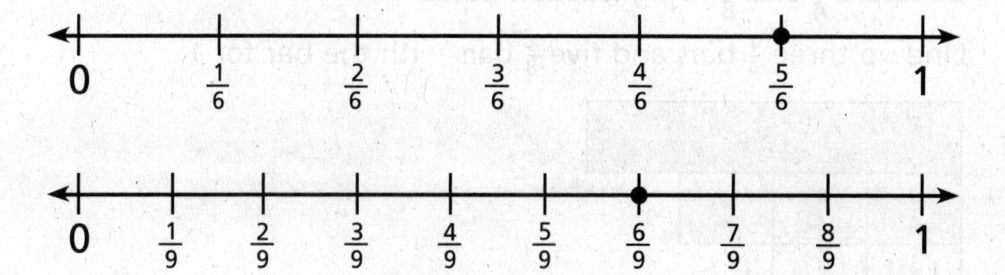

4. $\frac{2}{5} \bigcirc \frac{2}{3}$

5. $\frac{4}{9} \bigcirc \frac{1}{2}$

6. $\frac{4}{4} \bigcirc \frac{9}{9}$

7. $\frac{4}{6} \bigcirc \frac{2}{12}$

Check

8. Trevor says that you can tell $\frac{3}{9}$ is greater than $\frac{3}{7}$ because the numerators are the same and 9 is greater than 7. Is he correct? Explain why or why not.

_____
_____
_____
_____

IIN32   Response to Intervention • Tier 3

Name _____

Compare and Order Fractions
Skill 17

Learn the Math

Number lines can be used to compare and order fractions.

Compare $\frac{3}{4}$ **and** $\frac{7}{8}$.

Vocabulary
fraction
number line

Draw a number line divided into fourths. Then draw another number line below it, divided into eighths.

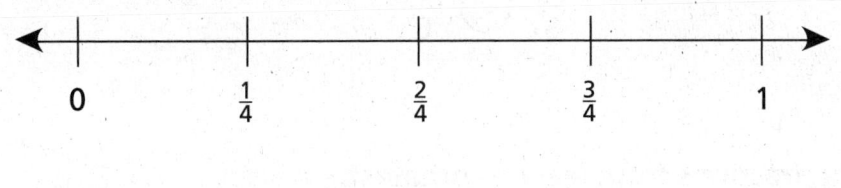

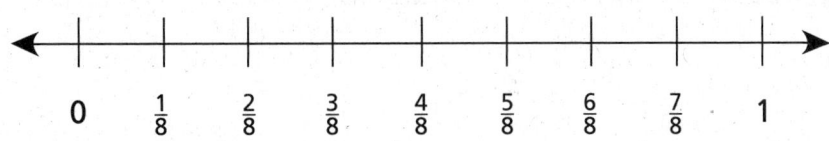

（number line showing 0, $\frac{1}{4}$, $\frac{2}{4}$, $\frac{3}{4}$, 1）

（number line showing 0, $\frac{1}{8}$, $\frac{2}{8}$, $\frac{3}{8}$, $\frac{4}{8}$, $\frac{5}{8}$, $\frac{6}{8}$, $\frac{7}{8}$, 1）

Find and mark the fractions on the lines. The one farther to the right is the greater fraction. So, $\frac{7}{8} > \frac{3}{4}$.

Order $\frac{13}{15}$, $\frac{4}{5}$, **and** $\frac{7}{10}$ **from least to greatest.**

Draw a number line divided into fifteenths, one divided into fifths, and another divided into tenths.

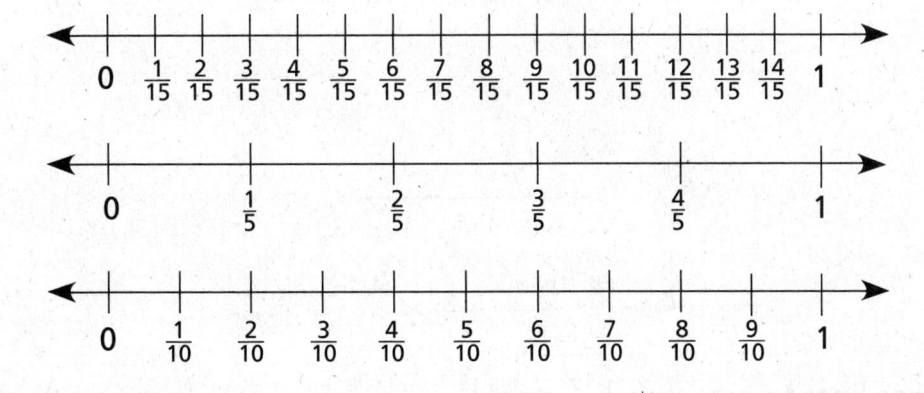

（number line showing 0, $\frac{1}{15}$ through $\frac{14}{15}$, 1）

（number line showing 0, $\frac{1}{5}$, $\frac{2}{5}$, $\frac{3}{5}$, $\frac{4}{5}$, 1）

（number line showing 0, $\frac{1}{10}$ through $\frac{9}{10}$, 1）

Find and mark the fractions on the number lines.

Order them from farthest left to farthest right.

So, the order from least to greatest is _____.

Response to Intervention • Tier 3   IIN33

Do the Math   Skill ⑰

Use number lines to compare. Write. <, >, or =.

1. $\frac{3}{8}$ ◯ $\frac{7}{16}$ 2. $\frac{5}{18}$ ◯ $\frac{2}{9}$

3. $\frac{9}{14}$ ◯ $\frac{6}{7}$ 4. $\frac{2}{3}$ ◯ $\frac{4}{6}$

Use number lines to order the fractions from least to greatest.

5. $\frac{2}{3}, \frac{3}{6}, \frac{5}{9}$ 6. $\frac{1}{2}, \frac{2}{5}, \frac{7}{10}$

_____   _____

Use number lines to order the fractions from greatest to least.

7. $\frac{3}{4}, \frac{7}{8}, \frac{5}{12}$ 8. $\frac{1}{3}, \frac{4}{6}, \frac{4}{9}$

_____   _____

Check

9. A student drew 3 number lines to compare $\frac{1}{3}$, $\frac{7}{9}$, and $\frac{11}{18}$. By what fractions did she divide each number line?

_____

_____

_____

IIN34   Response to Intervention • Tier 3

Name_____

Model Equivalent Fractions
Skill 18

Learn the Math

Equivalent fractions are fractions that name the same amount. You can use fraction bars and number lines to find equivalent fractions.

Vocabulary

equivalent fractions

Use fraction bars to find equivalent fractions for $\frac{1}{3}$.

Step 1

Use fraction bars. Line up one $\frac{1}{3}$ bar with the bar for 1 to show $\frac{1}{3}$.

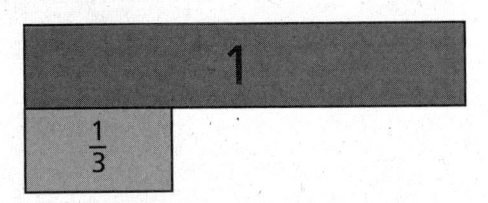

Step 2

Line up other bars of the same type to show the same amount as $\frac{1}{3}$.

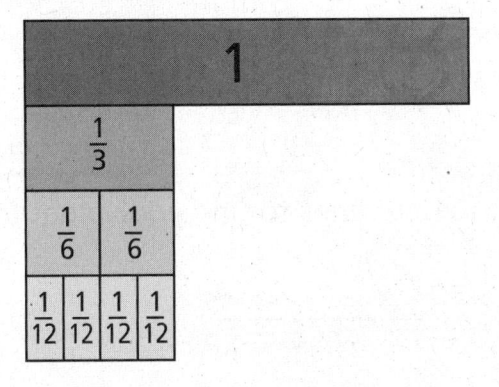

So, two fractions equivalent to $\frac{1}{3}$ are $\frac{2}{6}$ and $\frac{4}{12}$.

Use number lines to find equivalent fractions for $\frac{3}{4}$.

Step 1: Draw three number lines.

Divide one into fourths, one into eighths, and one into twelfths.

Step 2: Locate $\frac{3}{4}$ on the first number line.

Fractions that line up with $\frac{3}{4}$ are equivalent to $\frac{3}{4}$.

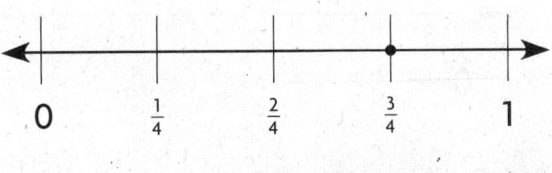

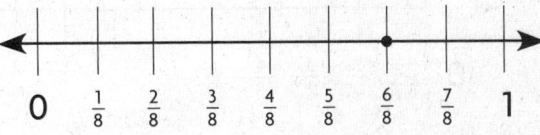

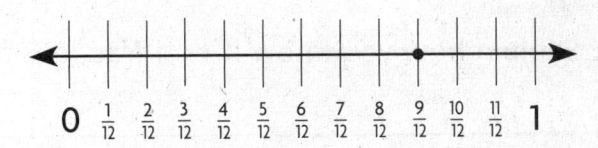

So, two fractions equivalent to $\frac{3}{4}$ are ____ and ____.

Response to Intervention • Tier 3   IIN35

Do the Math

Skill 18

Use fraction bars to find equivalent fractions.

1. $\frac{2}{3}$ = _____

2. $\frac{2}{4}$ = _____

Use number lines to find equivalent fractions.

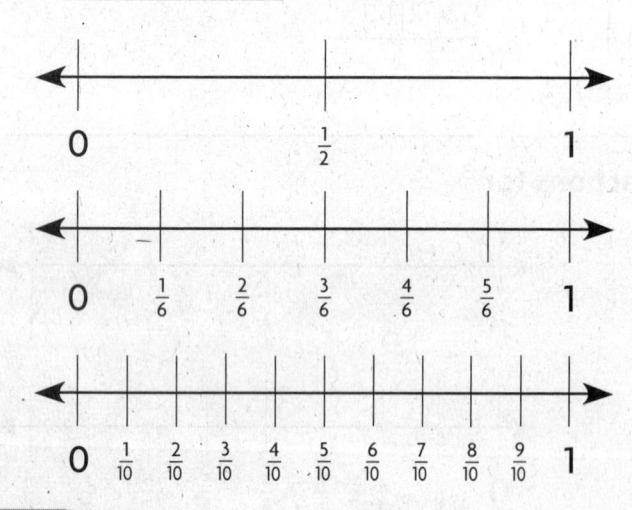

3. $\frac{1}{2}$ = _____

Check

4. Explain how you know if two fractions are equivalent.

_____

_____

_____

_____

IIN36 Response to Intervention • Tier 3

Name_____

Understand Mixed Numbers
Skill 19

Learn the Math

A mixed number is made up of a whole number and a fraction.

Use pattern blocks.

Model one and two sixths.

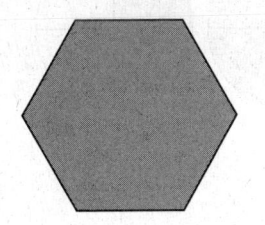

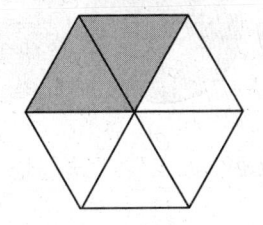

Vocabulary

improper fraction
mixed number

Think: one and two sixths is $1 + \frac{2}{6}$.

$$1 + \frac{2}{6} = 1\frac{2}{6}, \text{ or } 1\frac{1}{3}$$

Use a number line.

Draw a number line to locate $1\frac{3}{4}$ and $3\frac{1}{4}$.

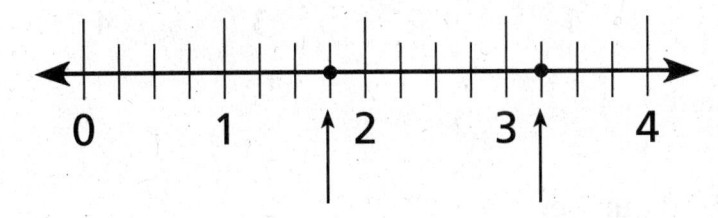

First, divide the number line into four equal parts.

Label the whole numbers.

Then, mark four equal parts between each whole number. Each part represents one fourth.

Locate and label $1\frac{3}{4}$ and $3\frac{1}{4}$.

Use fraction bars.

Rename $1\frac{3}{8}$ as a fraction.

First, use fraction bars to model $1\frac{3}{8}$.

Then, place $\frac{1}{8}$ bars under the bars for $1\frac{3}{8}$.

The total number of $\frac{1}{8}$ bars is the numerator of the fraction.

The numerator of the fraction is _____.

So, $1\frac{3}{8}$ renamed as a fraction is _____.

A fraction greater than 1 is sometimes called an improper fraction. So, $\frac{11}{8}$ is an improper fraction.

Response to Intervention • Tier 3   IIN37

Do the Math

Skill 19

Write a mixed number for each picture.

1. _____

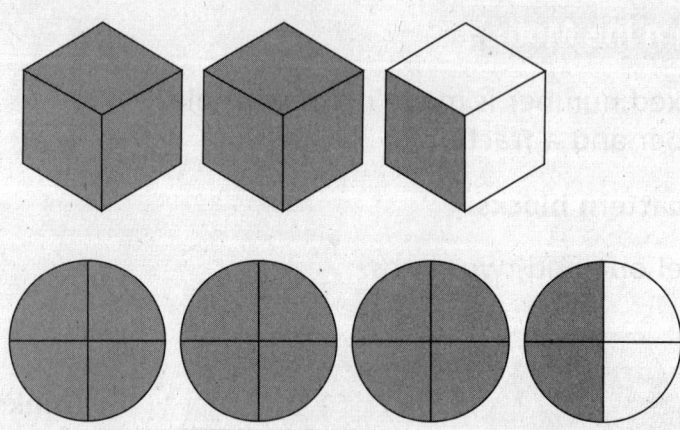

2. _____

Use the number line to write the mixed number represented by each point.

3. Point A _____

4. Point B _____

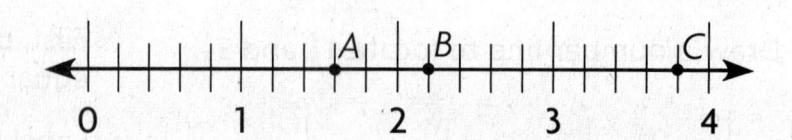

5. Point C _____

Use the fraction bars to rename the mixed number as an improper fraction.

6. $1\frac{3}{4}$ _____

7. $1\frac{1}{2}$ _____

Check

8. Ms. Jackson ordered three pizzas for her class. The class ate $2\frac{1}{2}$ pizzas. Draw a picture that represents the mixed number. Explain your drawing.

_____
_____
_____

IIN38   Response to Intervention • Tier 3

Name_____

Multiply or Divide to Find Equivalent Fractions

Skill 20

Learn the Math

Equivalent fractions are fractions that name the same amount. You can use multiplication and division to find equivalent fractions.

Use multiplication to find fractions that are equivalent to $\frac{2}{3}$.

Multiply both the numerator and denominator by any number except zero.

Vocabulary
denominator
equivalent fractions
numerator

Try 2. $\frac{2}{3} = \frac{2 \times 2}{3 \times 2} = \frac{4}{6}$

Think: Multiply the numerator and the denominator by the same number.

Try 4. $\frac{2}{3} = \frac{2 \times 4}{3 \times 4} = $ _____

So, $\frac{4}{6}$ and $\frac{8}{12}$ are both equivalent to $\frac{2}{3}$.

Use division to find fractions that are equivalent to $\frac{18}{36}$.

If the numerator and denominator have a common factor, you can divide by that factor.

Try 3. $\frac{18}{36} = \frac{18 \div 3}{36 \div 3} = \frac{6}{12}$

Think: Divide the numerator and the denominator by the same number.

Try 9. $\frac{18}{36} = \frac{18 \div 9}{36 \div 9} = $ _____

So, ____ and ____ are both equivalent to $\frac{18}{36}$.

Response to Intervention • Tier 3   IIN39

Do the Math

Skill 20

Multiply to find two equivalent fractions for each.

1. $\frac{2}{7}$ = _____
2. $\frac{1}{8}$ = _____
3. $\frac{3}{4}$ = _____
4. $\frac{1}{3}$ = _____
5. $\frac{7}{8}$ = _____
6. $\frac{1}{4}$ = _____

Divide to find two equivalent fractions for each.

7. $\frac{12}{24}$ = _____
8. $\frac{10}{30}$ = _____
9. $\frac{9}{36}$ = _____
10. $\frac{6}{6}$ = _____
11. $\frac{12}{20}$ = _____
12. $\frac{20}{24}$ = _____

Check

13. Are the fractions $\frac{4}{8}$ and $\frac{1}{2}$ equivalent? How do you know?

_____

_____

_____

IIN40 Response to Intervention • Tier 3

Name_____

Simplest Form
Skill ㉑

Learn the Math

A fraction is in simplest form when the only number that can be divided evenly into the numerator and the denominator is 1.

Vocabulary

simplest form

Use fraction bars to find the simplest form of $\frac{6}{9}$.

Step 1

Use fraction bars. Line up six $\frac{1}{9}$ bars with the bar for 1 to show $\frac{6}{9}$.

Step 2

Line up other fraction bars of the same type with denominators less than 9 to show the same amount as $\frac{6}{9}$.

Thirds are the largest fraction pieces that can be used.

Two $\frac{1}{3}$ bars show the same amount as six $\frac{1}{9}$ bars.

So, the simplest form of $\frac{6}{9}$ is _____.

Use division to find the simplest form of $\frac{18}{24}$.

Try 3. Divide the numerator and denominator by 3.

$$\frac{18}{24} = \frac{18 \div 3}{24 \div 3} = \frac{6}{8}$$

Next, try 2. Divide the numerator and denominator by 2.

$$\frac{6}{8} = \frac{6 \div 2}{8 \div 2} = \frac{3}{4}$$

Now the only number that can be divided into the numerator and denominator of $\frac{3}{4}$ is 1.

Think: The only common factor for 3 and 4 is 1.

So, the simplest form of $\frac{18}{24}$ is _____.

Do the Math

Skill 21

Use fraction bars to find the simplest form.

1. $\frac{9}{12}$ = _____

1
$\frac{1}{12}$ $\frac{1}{12}$ $\frac{1}{12}$ $\frac{1}{12}$ $\frac{1}{12}$ $\frac{1}{12}$ $\frac{1}{12}$ $\frac{1}{12}$ $\frac{1}{12}$
$\frac{1}{4}$ $\frac{1}{4}$ $\frac{1}{4}$

2. $\frac{8}{10}$ = _____

1
$\frac{1}{10}$ $\frac{1}{10}$ $\frac{1}{10}$ $\frac{1}{10}$ $\frac{1}{10}$ $\frac{1}{10}$ $\frac{1}{10}$ $\frac{1}{10}$
$\frac{1}{5}$ $\frac{1}{5}$ $\frac{1}{5}$ $\frac{1}{5}$

Is the fraction in simplest form? If not, write it in simplest form.

3. $\frac{8}{15}$ _____ 4. $\frac{12}{36}$ _____ 5. $\frac{10}{25}$ _____

6. $\frac{9}{54}$ _____ 7. $\frac{11}{12}$ _____ 8. $\frac{18}{40}$ _____

Check

9. Is the fraction $\frac{6}{24}$ in simplest form? How do you know?

_____

_____

_____

IIN42   Response to Intervention • Tier 3

Name_____

Add and Subtract Fractions

Skill 22

Learn the Math

Some fractions, like the fractions on this page, have like denominators. Fractions with like denominators have the same number in their denominators.

Add. $\frac{3}{5} + \frac{1}{5}$

Vocabulary

like fractions
simplest form

Shade 3 parts of a fifths model to show $\frac{3}{5}$.

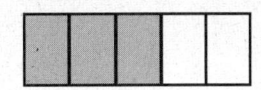

Then shade 1 more part to show $\frac{1}{5}$.

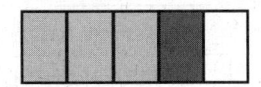

The shaded parts represent the sum.
Write the fraction for the parts that are shaded.

$$\frac{3}{5} + \frac{1}{5} = \frac{4}{5}$$

Check that the sum is in simplest form.

Subtract. $\frac{7}{9} - \frac{4}{9}$

Shade _____ parts of a ninths model to represent $\frac{7}{9}$.

To model subtracting $\frac{4}{9}$ from $\frac{7}{9}$, draw a line through _____ parts.

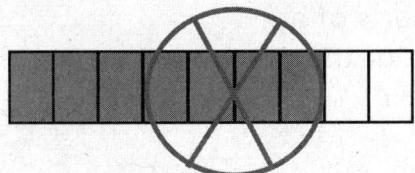

How many shaded parts do not have a line through them? ____

$\frac{7}{9} - \frac{4}{9} = \underline{}$

Write the difference in simplest form: $\frac{3}{9} = \frac{1}{3}$

Response to Intervention • Tier 3   IIN43

Do the Math

Shade the model to find the sum or difference. Write it in simplest form.

1. $\frac{1}{3} + \frac{2}{3} =$ _____

2. $\frac{6}{7} - \frac{5}{7} =$ _____

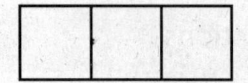

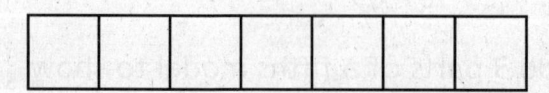

3. $\frac{5}{8} - \frac{2}{8} =$ _____

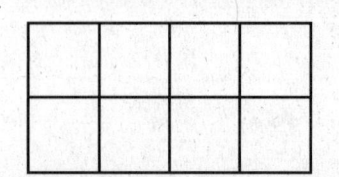

4. $\frac{4}{10} + \frac{4}{10} =$ _____

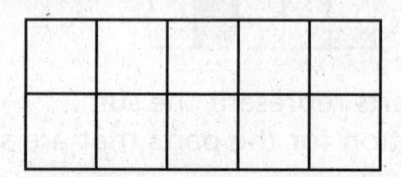

5. $\frac{1}{6} + \frac{1}{6} =$ _____

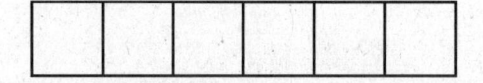

6. $\frac{9}{12} - \frac{2}{12} =$ _____

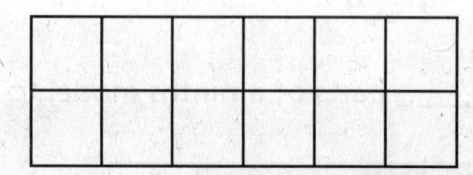

Check

7. A student was asked to add $\frac{3}{4} + \frac{1}{4}$. She shaded 3 parts of a fourths model. Then she crossed out 1 shaded part of the model. Did her model show the correct sum? Explain.

_____

_____

_____

Name _____

Rename Fractions and Mixed Numbers

Skill 23

Learn the Math

A mixed number is made up of a whole number and a fraction. There are several ways to rename mixed numbers as fractions, or to name fractions as mixed numbers.

Rename $1\frac{4}{5}$ as a fraction.

Use fraction bars.

Model $1\frac{4}{5}$.

Place $\frac{1}{5}$ bars under the bars for $1\frac{4}{5}$.

The total number of $\frac{1}{5}$ bars is the numerator of the fraction.

The numerator of the fraction is _____ .

Vocabulary

fraction
mixed number

So, $1\frac{4}{5}$ renamed as a fraction is _____ .

Use a number line.

First, divide the number line into 2 equal parts. Label the whole numbers.

Then mark 5 equal parts between each whole number. Each part represents one fifth.

Label the parts with mixed numbers and fractions.

Locate $1\frac{4}{5}$.

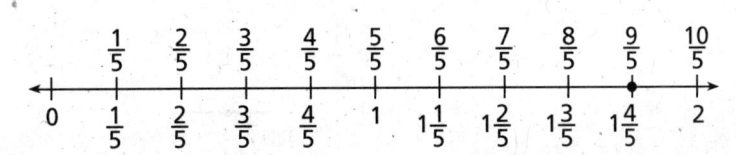

So, $1\frac{4}{5}$ renamed as a fraction is _____ .

Rename $\frac{13}{4}$ as a mixed number.

Since $\frac{13}{4}$ means $13 \div 4$, you can use division to rename a fraction greater than 1 as a mixed number.

Write the quotient, 3, as the whole number part.

Then write the remainder, 1, as the numerator and the divisor, 4, as the denominator.

denominator → $4\overline{)13}$ ← numerator
$\underline{-12}$
1 ← number of fourths left over

So, $\frac{13}{4}$ renamed as a mixed number is _____ .

Response to Intervention • Tier 3   IIN45

Do the Math

Skill 23

Use the number line to rename each fraction as a mixed number and each mixed number as a fraction.

1. $5\frac{2}{5} =$ _____

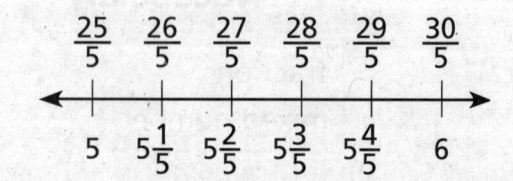

2. $\frac{29}{5} =$ _____

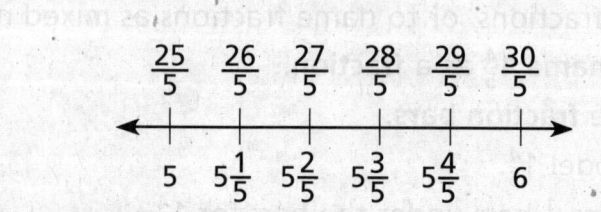

Rename each fraction as a mixed number.

3. $\frac{28}{3} =$ _____

4. $\frac{56}{9} =$ _____

5. $\frac{26}{7} =$ _____

6. $\frac{44}{9} =$ _____

Rename the mixed number as a fraction.

7. $3\frac{1}{6} =$ _____

8. $4\frac{2}{7} =$ _____

9. $4\frac{2}{5} =$ _____

10. $5\frac{5}{8} =$ _____

Check

11. Paula used division to rename $\frac{35}{8}$ as a mixed number. Her work is shown at the right. What did she do wrong?

4
$8\overline{)35}$
$\underline{-32}$
3

$\frac{35}{8} = 4\frac{3}{4}$

_____

_____

_____

IIN46 Response to Intervention • Tier 3

Name_____

Relate Fractions and Decimals
Skill 24

Learn the Math

A decimal is a number with one or more digits to the right of the decimal point. Like fractions, decimals show values less than 1.

Vocabulary
- decimal
- decimal point
- fraction
- hundredths
- tenths

Use models to read and write fractions as decimals.

	Fraction	Decimal
Shade the model to show 1.	**Read:** one **Write:** $\frac{1}{1}$	**Read:** one **Write:** 1.0
Shade $\frac{4}{10}$ of the model.	**Read:** four tenths **Write:** _____	**Read:** _____ **Write:** 0.4
Shade $\frac{5}{100}$ of the model.	**Read:** five hundredths **Write:** _____	**Read:** _____ **Write:** 0.05

Use a number line to show the same amount.

A number line divided into 100 equal parts can be used to model fractions and decimals.

Locate the point 0.45.

What fraction names this point on the number line? _____

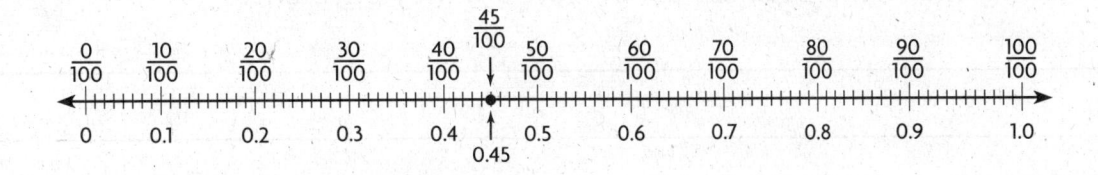

[Number line showing $\frac{0}{100}$ to $\frac{100}{100}$ and 0 to 1.0, with $\frac{45}{100}$ marked at 0.45]

So, 0.45 names the same amount as _____ or $\frac{9}{20}$.

Response to Intervention • Tier 3   IIN47

Do the Math

Skill 24

Write the fraction and decimal for each model.

1.

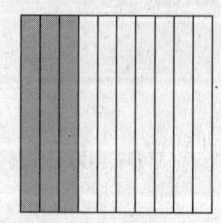

decimal: _____   fraction: _____

2.

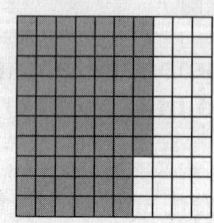

decimal: _____   fraction: _____

3.

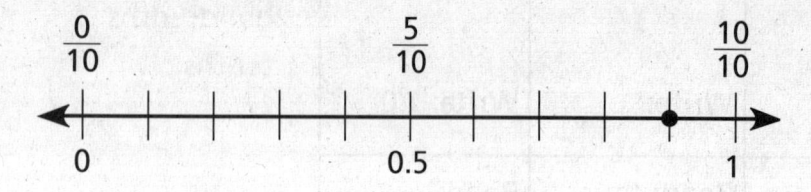

decimal: _____   fraction: _____

Write each fraction as a decimal. Draw a picture or make a model to help you.

4. $\frac{8}{100}$ _____

5. $\frac{3}{4}$ _____

6. $\frac{6}{10}$ _____

7. $\frac{52}{100}$ _____

Check

8. How are fractions and decimals alike? How are they different?

_____

_____

_____

_____

Name_____

Decimal Models
Skill 25

Learn the Math

A decimal uses place value to show values less than one. Understanding fractions that have a denominator of 10 or 100 will help you understand decimals.

Vocabulary
decimal
hundredths
tenths

Model 0.7 using a decimal model.

ONES	.	TENTHS
0	.	7

The number to the right of the decimal point is in the tenths place.

decimal point ↑

Step 1	Use a model with 10 equal parts. Each equal part is one tenth, or $\frac{1}{10}$.
Step 2	The number after the decimal point is **7**. So, shade **7** of the **10** equal parts.
Step 3	Write the decimal for the shaded amount. _____ Write the fraction for the shaded amount. _____ The decimal and the fraction name the same amount.

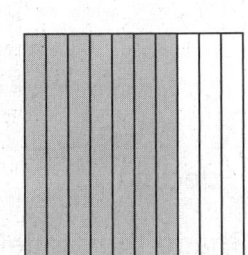

Model 0.53 using a decimal model.

ONES	.	TENTHS	HUNDREDTHS
0	.	5	3

The first number to the right of the decimal point is in the tenths place. The second number is in the hundredths place.

Step 1	Use a model with 100 equal parts. Each equal part is one hundredth, or $\frac{1}{100}$.
Step 2	The number after the decimal point is **53**. So, shade **53** of the **100** equal parts.
Step 3	Write the decimal for the amount you shaded. _____ Write the fraction for the amount you shaded. _____ The decimal and the fraction name the same amount.

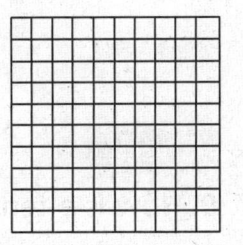

Response to Intervention • Tier 3 IIN49

Do the Math　　　　　　　　　　　　　　　　　　　　　　　　　　　Skill 25

Write the decimal and the fraction for each shaded part.

1.

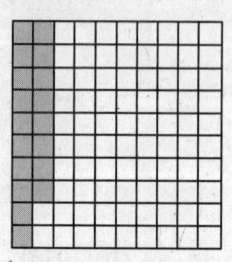

decimal: _____   fraction: _____

2.

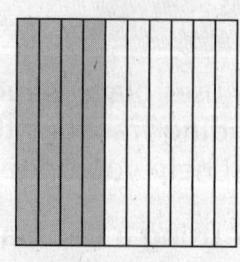

decimal: _____   fraction: _____

3.

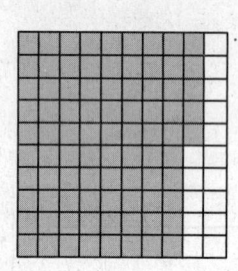

decimal: _____   fraction: _____

4.

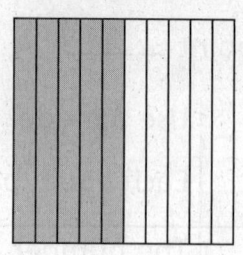

decimal: _____   fraction: _____

Shade the model to show the decimal.

5. 0.27

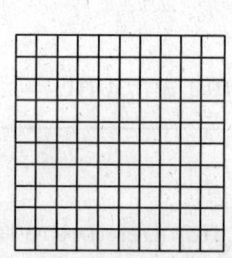

6. 0.1

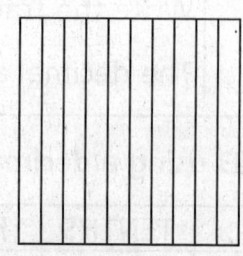

7. 0.9

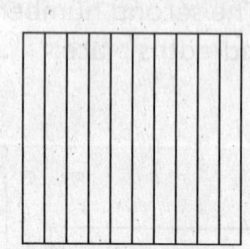

8. 0.30

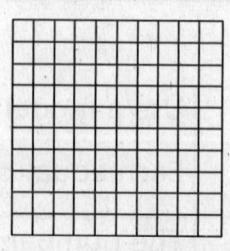

Check

9. How is a tenths model different from a hundredths model?

_____

_____

_____

IIN50   Response to Intervention • Tier 3

Name_____

Compare Decimals
Skill 26

Learn the Math

Use models or number lines to compare decimals.

Compare 1.4 and 1.7 using models.

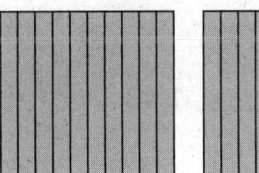

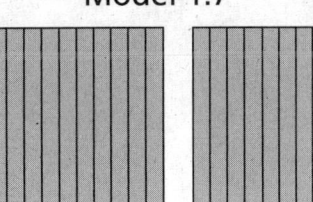

Model 1.4   Model 1.7

Vocabulary

decimal
number line

Which model has more shaded parts?_____

So, 1.7 > 1.4, 1.7 is greater than 1.4.

Or, 1.4 < 1.7, 1.4 is less than 1.7.

Compare 3.39 and 3.3 using a number line.

Write an equivalent decimal for 3.3.   3.3 = 3.30

Find the numbers on the number line:

```
3.10      3.20      3.30      3.40
```

The greater decimal is farther to the right.

Which decimal is farther to the right? _____

So _____ > _____.

Compare 2.61 and 2.74 using models.

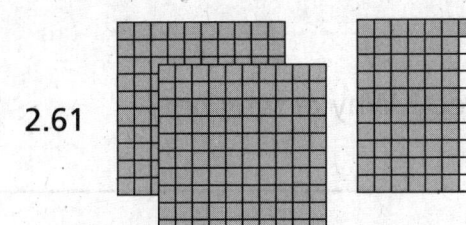

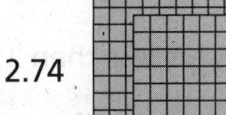

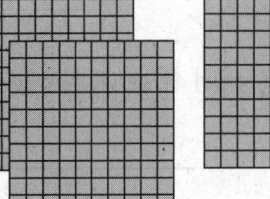

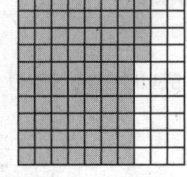

2.61   2.74

Which model has more shaded parts? _____

So, _____ > _____, or _____ < _____.

Response to Intervention • Tier 3   IIN51

Do the Math

Skill 26

Compare the decimals. Write <, >, or =.

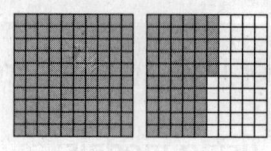

1. 0.6   0.9

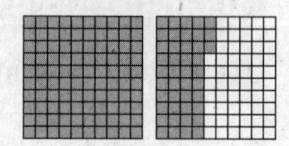

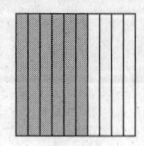

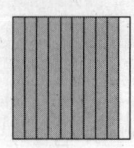

   0.6 _____ 0.9

2. 1.55   1.43

   1.55 _____ 1.43

3. 2.29 and 2.33

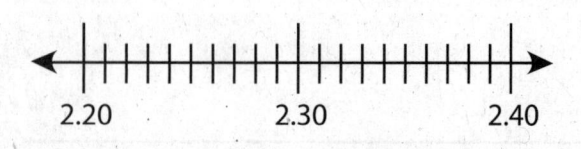

   2.29 _____ 2.33

4. 5.8 and 5.78

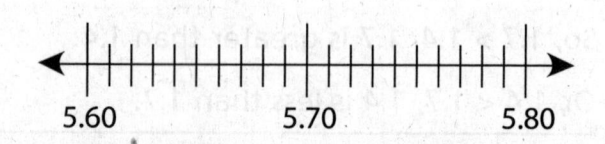

   5.8 _____ 5.78

5. 1.12 and 1.02

   1.12 _____ 1.02

6. 1.67 and 1.76

   1.67 _____ 1.76

7. 1.46 and 1.64

   1.46 _____ 1.64

8. 1.34 and 1.34

   1.34 _____ 1.34

Check

9. A student said that 1.1 is less than 1.10. Is he correct? Why or why not?

_____

_____

_____

IIN52   Response to Intervention • Tier 3

Name_____

Multiply Decimals by Whole Numbers

Skill 27

Learn the Math

Decimals can be multiplied in the same way as whole numbers.

Find the product. Use estimation.

2.5 × 4

Vocabulary

decimal

Step 1
Estimate the product. Round each factor.

2.5 × 4
↓   ↓
3 × 4 = 12

Step 2
Multiply as with whole numbers.

$$\begin{array}{r} \overset{2}{}2.5 \\ \times\ 4 \\ \hline 100 \end{array}$$

Step 3
Use the estimate to place the decimal point in the product.

$$\begin{array}{r} \overset{2}{}2.5 \\ \times\ 4 \\ \hline 10.0 \end{array}$$

↑
The product should have the closest place value to the estimate.

So, 2.5 × 4 is _____.

Find 3.46 × 8. Count the decimal places.

Step 1 Multiply as with whole numbers

Step 2 Count the number of decimal places in both factors. Place the decimal point that number of places from the right in the product.

$$\begin{array}{r} \overset{3\ 4}{}3.46 \\ \times\ 8 \\ \hline 27.68 \end{array}$$

← 2 decimal places in 3.46
← 0 decimal places in 8
← 2 + 0 decimal places in the product

Place the decimal point ____ places from the right in the product.

So, 3.46 × 8 is _____.

Find 5.2 × 19. Count the decimal places.

Step 1 Multiply as with whole numbers.

Step 2 Count the number of decimal places in both factors.

$$\begin{array}{r} \overset{1}{}5.2 \\ \times\ 19 \\ \hline 468 \\ 520 \\ \hline 98.8 \end{array}$$

← 1 decimal place in 5.2
← 0 decimal places in 19

← 1 + 0 decimal places in the product

Place the decimal point ____ place from the right in the product.

So, 5.2 × 19 is _____.

Response to Intervention • Tier 3   IIN53

Do the Math   Skill **27**

Find the product. Show your work.

1. 2.77   2. 1.6
    × 5   × 21

3. 2.3   4. 1.44
    × 16   × 5

5. 0.92   6. 4.4
    × 7   × 12

7. 1.5   8. 0.63
    × 22   × 19

Check

9. Josh says that 1.32 × 3 is 39.6. Is he correct? Explain.

_____

_____

_____

Name_____

Understand Percent
Skill 28

Learn the Math

You can write a fraction or a decimal as a percent.

Percent, or %, means "per hundred."

$100\% = \frac{100}{100} = \frac{1}{1} = 1$

The model shows 100%. →
100 squares of 100 are shaded.

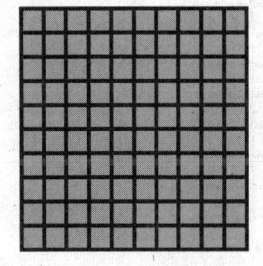

Vocabulary

percent

$50\% = \frac{50}{100} = \frac{1}{2} = 0.50$

Shade the model to show 50% by shading half of the boxes.

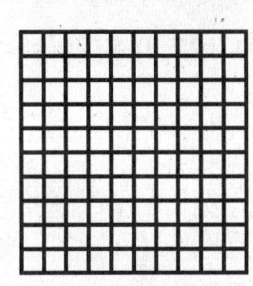

Write a percent as a fraction in simplest form.

25%

Step 1 Write the percent as a fraction with a denominator of 100.

$25\% = \frac{25}{100}$

Step 2 Write the fraction in simplest form.

$\frac{25}{100} = \frac{25 \div 25}{100 \div 25} = \frac{1}{4}$

So, 25% written as fraction in simplest form is _____ .

Write a percent as a decimal.

35%

Step 1 Write the percent as a fraction with a denominator of 100.

$35\% = \frac{35}{100}$

Step 2 Write the fraction as a decimal.

$\frac{35}{100} = 0.35$

So, 35% written as a decimal is _____ .

Response to Intervention • Tier 3   IIN55

Do the Math

Shade in the model to show the percent.

1. 75%

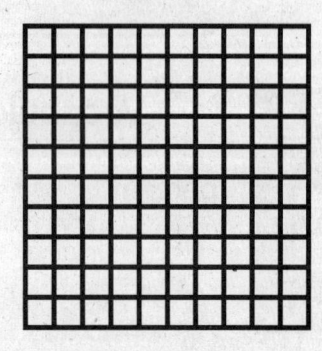

2. 34%

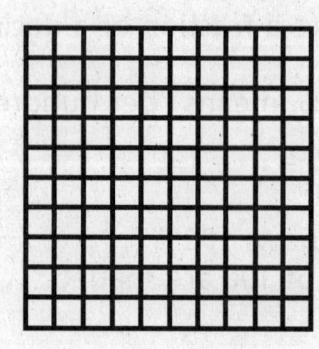

Write each percent as a fraction in simplest form.

3. 55%

4. 91%

_____

_____

Write each percent as a decimal.

5. 25%

6. 5%

_____

_____

7. 10%

8. 99%

_____

_____

Check

9. Lawrence said that 60% of the students in his class voted to play kickball at recess. He says that $\frac{6}{10}$ shows this percent as a fraction in simplest form. Do you agree? Explain.

_____

_____

_____

Name_____

Read a Frequency Table
Skill 29

Learn the Math

Frequency tables show information that can be measured or counted.

Jim asked each student in his class to name their favorite pet and recorded the answers. First, he made a tally table to show the results of his survey.

Vocabulary
frequency
numerical data
survey

Favorite Pet Survey	
Pet	Tally
dog	||||| |||||
cat	||||| ||||| |||
hamster	||
snake	|||

Then Jim made a frequency table. It shows numerical data.

Favorite Pet Survey	
Pet	Frequency
dog	10
cat	13
hamster	2
snake	3

One column shows the pets students chose.

The other column shows the number of times each pet was chosen.

Frequency is the number of times something was chosen, or occurred.

Use the frequency table to answer each question.

How many students did Jim survey? _____

Which pet was chosen most often? _____

How many more children chose cats than snakes? _____

Response to Intervention • Tier 3   IIN57

Do the Math Skill 29

Use the frequency table at the right for Problems 1–3.

1. How many more players are on the football team than the basketball team?

   _____

2. Which team has the least number of players?

   _____

3. How many players are on the hockey team?

   _____

School Sports Teams	
Team	Frequency (Number of Players)
hockey	20
football	41
basketball	19
baseball	20

Use the frequency table below for Problems 4–6.

4. Is this statement true or false: "More people voted for drama than for comedy"?

   _____

5. How many people voted in all?

   _____

6. Which type of movie got 40 votes?

   _____

Favorite Type of Movie	
Type of Movie	Frequency (Votes)
comedy	35
horror	27
drama	33
action	40

Check

7. Use the Movie frequency table above. Order the types of movie from the type with the greatest number of votes to the type with the least number of votes. Explain how you found the answer.

   _____

   _____

   _____

IIN58 Response to Intervention • Tier 3

Name_____

Mean

Skill 30

Learn the Math

The mean is the average of a set of numbers. You can find the mean by using models or by using addition and division.

Vocabulary

mean

Use models to find the mean of these numbers.

2, 7, 6, 9

Step 1 Model each number using stacks of cubes.

2   7   6   9

Step 2 Rearrange the stacks so that each has the same number of cubes.

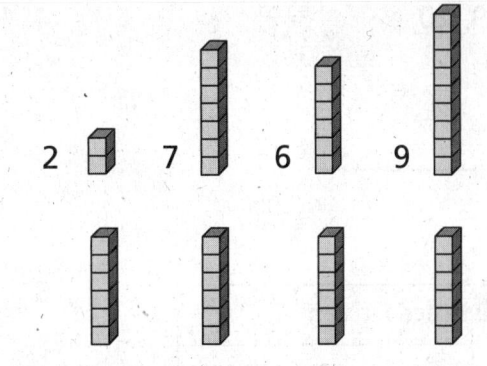

The number of cubes in each of the equal stacks is the mean. There are __6__ cubes in each stack, so the mean is ____.

Use addition and division to find the mean.

Find the mean number of miles biked.

Step 1 Add all of the numbers in the data set.
4 + 9 + 14 + 8 + 5 = ____
The sum is ____.

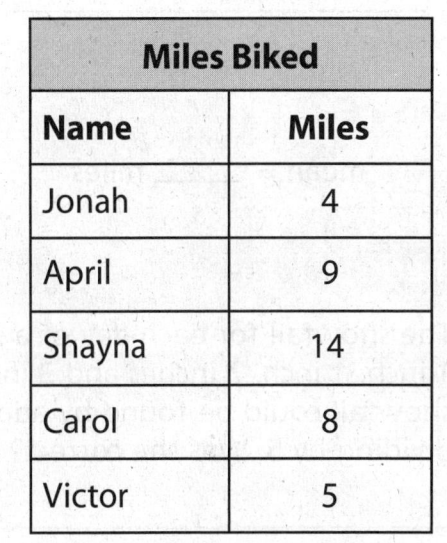

Step 2 Count the number of addends.

There are ____ numbers in the set.

So, there are ____ addends.

Step 3 Divide the sum by the number of addends.

____ ÷ ____ = ____

So, the mean number of miles biked is ____ miles.

Miles Biked	
Name	Miles
Jonah	4
April	9
Shayna	14
Carol	8
Victor	5

Response to Intervention • Tier 3   IIN59

Do the Math

Skill 30

Find the mean of each set of data.

1. 30, 40, 50, 36

2. 1, 3, 4, 4, 5, 1

_____

_____

3. 23, 12, 16, 19, 20

4. 10, 11, 8, 6, 5

_____

_____

5.

Distance Driven	
Day	Distance (in miles)
Monday	99
Tuesday	91
Wednesday	92
Thursday	94

mean = _____ miles

6.

Baseball Games Attended	
Month	Number of Games
May	10
June	14
July	6
August	6
September	9

mean = _____ games

Check

7. The snowfall for each day of a school week was: 3 inches, 1 inch, 1 inch, 7 inches and 3 inches. Lisa said that the mean snowfall could be found by adding 3 + 1 + 7 inches, and then dividing by 5. Was she correct? Explain.

_____

_____

_____

IIN60   Response to Intervention • Tier 3

Name_____

Read a Pictograph
Skill 31

Learn the Math

Pictographs show information using pictures or symbols.

The Weather Service records the amount of snowfall in Jamestown. The pictograph shows the amount of snowfall for 5 different months.

Vocabulary
key
pictograph

Jamestown Snowfall	
December	❄❄
January	❄❄❄❄
February	❄❄❄
March	❄❄
April	❄

Key: Each ❄ = 2 inches of snow.

This pictograph uses ❄ as the symbol.

The key tells the amount that each symbol represents.

Each ❄ represents 2 inches of snow.

So, _____ inches fell in January.

The pictograph below shows the results of a favorite season survey.

Favorite Season Survey	
Season	Votes
Spring	☺☺☺☺☾
Summer	☺☺☺☺☺☺
Autumn	☺☺☺☺☾
Winter	☺☺☺

Key: Each ☺ = 6 votes.

This pictograph uses ☺ as the symbol.

The key shows that each ☺ represents 6 votes.

So, ☾ represents _____ votes.

So, Spring got _____ votes.

Response to Intervention • Tier 3   IIN61

Do the Math Skill 31

Use the Concert Attendance pictograph for Problems 1–3.

1. How many more people attended a concert on Saturday than Friday?

   _____

2. On which day did 60 people attend a concert?

   _____

3. How many people attended on Thursday?

   _____

Concert Attendance	
Tuesday	♪
Wednesday	♪♪♪
Thursday	♪♪♪
Friday	♪♪♪♪
Saturday	♪♪♪♪♪

Key: Each ♪ = 20 people.

Use the pictograph below for Problems 4–6.

4. What was the total number of people who were surveyed?

   _____

5. Which snack got the fewest votes?

   _____

6. How many votes did popcorn get?

   _____

Favorite Snack Survey	
Popcorn	♥♥♥♥
Apples	♥♥♥♥◗
Raisins	♥♥♥◗
Nuts	♥♥

Key: Each ♥ = 8 votes.

Check

7. Dave read the Favorite Snack Survey pictograph above, and said that popcorn and apples got the same number of votes. Is he correct? Explain.

_____

_____

IIN62   Response to Intervention • Tier 3

Name_____

Read a Bar Graph
Skill 32

Learn the Math

Bar graphs display data. A bar graph can use horizontal or vertical bars to show data.

This bar graph shows the number of flowers a store sold.

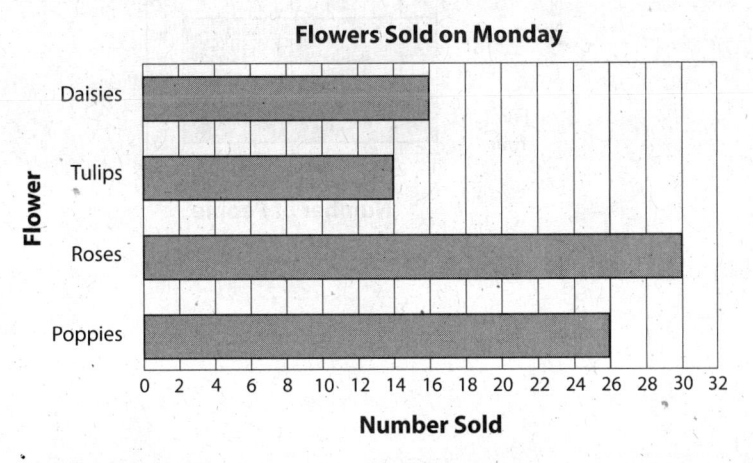

Flowers Sold on Monday

(Horizontal bar graph showing Flower on y-axis: Daisies, Tulips, Roses, Poppies; Number Sold on x-axis: 0 to 32)

Vocabulary
- bar graph
- range

Which type of flower was sold the most?

Think: Which bar is the longest? _____

How many daisies were sold?

Find the daisies bar. Follow the end of the bar down to the scale.
The number it matches is _____ .

The bar graph below shows the number of students in 4 different school clubs.

What is the range of the number of students in these clubs?

Find the tallest bar. It belongs to the Film Club.
There are _____ students in this club.

Find the shortest bar. It belongs to the Latin Club.
The bar falls halfway between 20 and 30.
So, there are _____ students in this club.

Students in School Clubs

(Vertical bar graph: Theater, Chess, Latin, Film on x-axis; Number of Students 0–80 on y-axis)

Subtract the least value from the greatest value.

_____ − _____ = _____

So, the range is _____ .

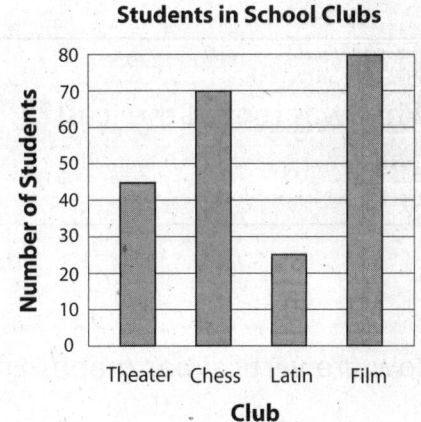

Response to Intervention • Tier 3 IIN63

Do the Math

Skill 32

Use the bar graph at the right for Problems 1–3.

1. Which two vacinations were each chosen 14 times?

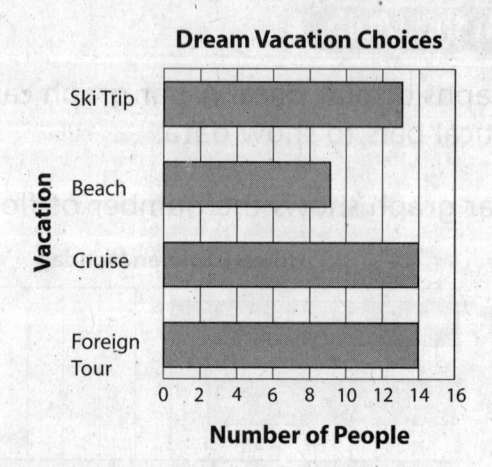

   _____

2. How many people chose ski trip?

   _____

3. How many more people chose cruise than beach?

   _____

Dream Vacation Choices

Use the bar graph at the right for Problems 4–6.

4. What is the range of the number of items rented?

   _____

5. How many bats were rented?

   _____

6. What was rented the least?

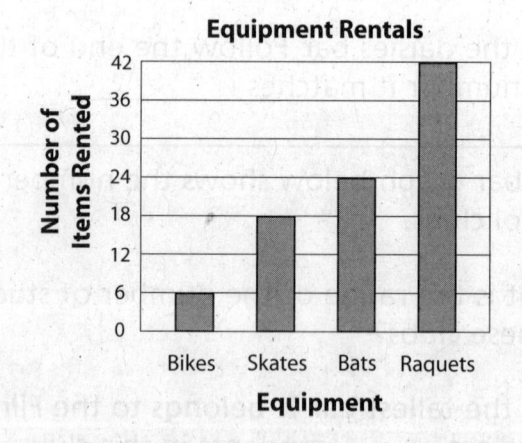

   _____

Equipment Rentals

Check

7. How are vertical bar graphs and horizontal bar graphs different?

   _____
   _____
   _____

IIN64   Response to Intervention • Tier 3

Name_____

Circle Graphs
Skill 33

Learn the Math

Circle graphs show data as parts of a whole.

Randall asked 16 of his friends to vote for their favorite class. Their responses are shown in the table below.

Class	English	Art	Drama	Math
Votes	1	4	8	3

Vocabulary

circle graph

Randall made a circle graph to show the data. First, he divided a circle into 16 equal parts. Each part represents one friend's response.

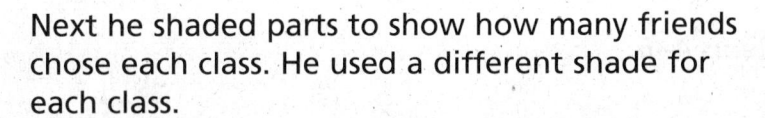

Next he shaded parts to show how many friends chose each class. He used a different shade for each class.

The greatest part shows the class chosen by the most friends.

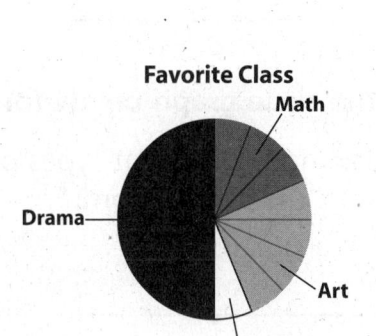

Favorite Class

This circle graph shows the time Nicole spent on her homework.

Which subject did Nicole spend the most time on?

Think: Which subject takes up the greatest part of the circle?

Nicole spent the most time on _____.

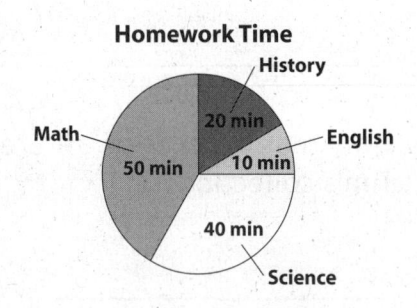

Homework Time

What is the total time Nicole spent studying?

To find out, add the times for all the subjects.

50 + 40 + 10 + 20 = _____ minutes

Nicole spent _____ minutes or _____ hours studying.

Response to Intervention • Tier 3   IIN65

Do the Math Skill 33

Use the circle graph at the right for Problems 1–3.

1. Who received the most votes?

   _____

2. Who received the fewest votes?

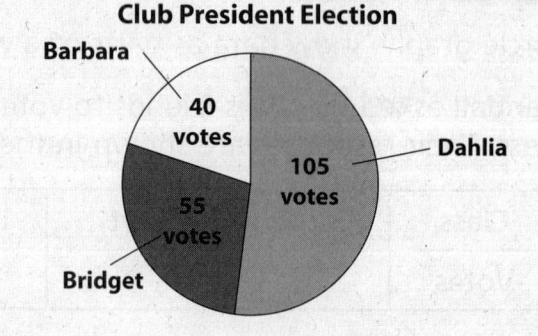

   _____

3. Who received more votes than Bridget?

   _____

Club President Election
- Barbara: 40 votes
- Dahlia: 105 votes
- Bridget: 55 votes

Use the circle graph below for Problems 4–6.

4. How many different types of art are in Tim's collection?

   _____

5. Are there more drawings than photos in Tim's collection?

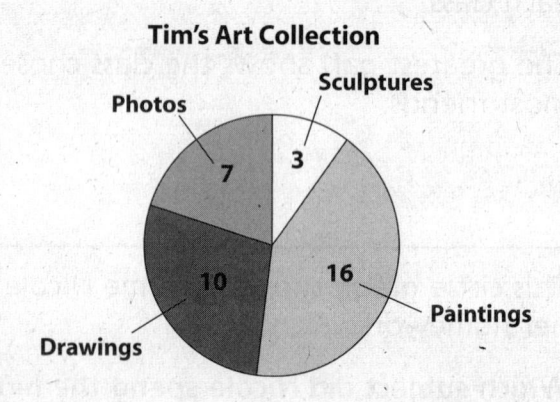

   _____

6. How many pieces of art are in Tim's collection?

   _____

Tim's Art Collection
- Photos: 7
- Sculptures: 3
- Paintings: 16
- Drawings: 10

Check

7. Judi read the Art Collection circle graph. She said that paintings represent $\frac{1}{2}$ of Tim's art collection. Was she correct? Explain.

   _____
   _____
   _____

IIN66 Response to Intervention • Tier 3

Name_____

Addition Properties
Skill 34

Learn the Math

Addition properties can help simplify problems.

Identity (or Zero) Property of Addition

The sum of any number and zero is equal to that number.

16 + 0 = 16

3,946 + 0 = _____

0 + 88 = ____

Vocabulary

Associative Property of Addition

Commutative Property of Addition

Identity (Zero) Property of Addition

Commutative Property of Addition

The numbers in an addition problem can be added in any order. The sum will not change.

| 26 + 10 = 10 + 26 | 9 + 80 = ____ + 9 |
| 36 = 36 | 89 = ____ |

Associative Property of Addition

Addends in an addition problem can be grouped in any order. The way the addends are grouped does not change the sum.

(9 + 7) + 3 = 9 + (7 + 3)	49 + (51 + 27) = (____ + 51) + 27
16 + 3 = 9 + 10	49 + 78 = ____ + 27
19 = 19	127 = ____

When using the Associative Property to solve problems, try to group numbers that are easy to add mentally.

Response to Intervention • Tier 3   IIN67

Do the Math

Skill 34

Find the missing number. Write which addition property you used.

1. 165 + ____ = 165

2. (44 + ____) + 2 = 44 + (16 + 2)

_____

_____

3. (57 + 3) + 22 = 57 + (____ + 22)

4. 13 + 39 = 39 + ____

_____

_____

5. 28 + (3 + 17) = (____ + 3) + 17

6. 1,399 + 0 = _____

_____

_____

7. 12 + (17 + ____) = (12 + 17) + 18

8. ____ + 0 = 11

_____

_____

9. (17 + 6) + ____ = 17 + (6 + 24)

10. 32 + ____ = 14 + 32

_____

_____

Check

11. How would you use the Associative Property to group these numbers so that you can add mentally? What is the sum?

33 + 5 + 115

_____

_____

_____

IIN68   Response to Intervention • Tier 3

Name_____

Multiplication Properties
Skill 35

Learn the Math

Multiplication properties can help you solve problems.

Zero Property of Multiplication

The product of any number and zero is zero.

14 × 0 = 0   0 × 333 = 0   64 × 0 × 65 = ____

Vocabulary

Associative Property of Multiplication

Commutative Property of Multiplication

Identity Property of Multiplication

Zero Property of Multiplication

Identity Property of Multiplication

The product of any number and one is that number.

17 × 1 = 17   1 × 63 = 63   4,955 × 1 = _____

Commutative Property of Multiplication

Two factors can be multiplied in any order without changing the product.

2 × 10 = 10 × 2   9 × 8 = 8 × 9   542 × 4 = 4 × _____

20 = 20   72 = 72   2,168 = _____

Associative Property of Multiplication

When there are three or more factors in a multiplication problem, they can be grouped in any order. The way the factors are grouped does not change the product.

(4 × 5) × 10 = 4 × (5 × 10)   (8 × 3) × 3 = 8 × (3 × 3)

20 × 10 = 4 × 50   24 × 3 = ____ × 9

200 = 200   72 = ____

When using the Associative Property to solve problems, try to group numbers that are easy to add mentally.

Response to Intervention • Tier 3   IIN69

Do the Math

Skill 35

Find the missing number. Write which multiplication property you used.

1. 12 × ____ = 12

   _____

2. 15 × ____ = 3 × 15

   _____

3. 3 × 4 × 8 × 0 = ____

   _____

4. (1 × 3) × 12 = 1 × (3 × ____)

   _____

5. 7 × 6 = ____ × 7

   _____

6. 199 × 0 = ____

   _____

7. 8 × 4 = ____ × 8

   _____

8. 17 × ____ = 17

   _____

9. (4 × 5) × ____ = 4 × (5 × 7)

   _____

10. ____ × 1 = 14

    _____

11. 8 × 7 = ____ × 8

    _____

12. 11 × ____ = 0

    _____

Check

13. How would you use the Commutative Property to find an equivalent expression for 14 × 2? What is the product?

    _____

    _____

    _____

IIN70   Response to Intervention • Tier 3

Name_____

Expressions
Skill 36

Learn the Math

You can write an expression to help you solve a problem. An expression has operation signs, numbers, and sometimes variables, but no equal sign.

Vocabulary
expression
variable

Walter buys three apples, then he buys four pears.

Write an expression to describe the amount of fruit Walter buys.

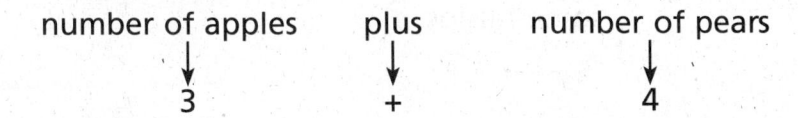

number of apples	plus	number of pears
↓	↓	↓
3	+	4

Find the value of the expression to find the amount of fruit Walter buys.

The value of 3 + 4 is __7__.

So, Walter bought _____ pieces of fruit.

Jay had 52 paintings. He gave some away to friends.

Write an expression to show how many paintings Jay has left.

Use a variable to represent the number of paintings Jay gave away since this value is not known.

original number of paintings	minus	number of paintings Jay gave away
↓	↓	↓
52	−	p

Suppose Jay gave away 15 paintings.

Find the value of the expression to find the number of paintings Jay has left.

52 − p Replace p with 15.
↓
52 − 15   Subtract 15 from 52.
↓
37

So, Jay has _____ paintings left.

Response to Intervention • Tier 3 IIN71

Do the Math

Skill 36

Write an expression to match the words.

1. Rosie has 12 pairs of socks. She buys 8 more pairs.

   _____

2. Simon had 13 library books. He returned 7 to the library.

   _____

Write an expression with a variable. Tell what the variable represents.

3. A girl had a bunch of balloons. Two of the balloons popped.

   _____
   _____

4. Jillian ate 9 dried apricots and some raisins.

   _____
   _____

Find the value of the expression. Show your work.

5. $33 - t$ if $t = 17$

   _____
   _____

6. $100 + g$ if $g = 101$

   _____
   _____

7. $103 - b$ if $b = 23$

   _____
   _____

8. $13 + f$ if $f = 100$

   _____
   _____

Check

9. Georgia baked 5 pies and gave a few away. The expression $5 - p$ shows how many pies she has left. What is the variable in this expression? What does the variable represent?

   _____
   _____

IIN72   Response to Intervention • Tier 3

Name_____

Number Patterns
Skill 37

Learn the Math

Number patterns can be explained by rules.

Write a rule for the pattern. Then find the missing numbers in the pattern.

3, 6, 9, 12, 15, ☐, ☐, 24

Step 1 Find a rule. **Think:** What rule changes 3 to 6? Try *multiply by 2*.
Test the rule: 3 × 2 = 6; 6 × 2 = _____, 6 × 2 ≠ 9
The rule *multiply by 2* does not work.

Try another rule. **Think:** What other rule changes 3 to 6? Try *add 3*.
Test the rule:

+3 +3 +3 +3 +3 +3 +3
3, 6, 9, 12, 15, ☐, ☐, 24

The rule *add 3* works.

Step 2 Use the rule to find the missing numbers.   15 + 3 = _____   18 + 3 = _____

So, the missing numbers are _____ and _____.

Some rules have more than one operation. Try two operations to find a rule for a pattern in which numbers decrease and then increase.

−4 ×2 −4 ×2 −4 ×2
14, 10, 20, 16, 32, ☐, ☐

The rule for this pattern is *subtract 4, multiply by 2*.

Use the rule to extend the pattern.

32 − 4 = _____   28 × 2 = _____

So, the next two numbers are _____ and _____.

Response to Intervention • Tier 3 IIN73

Do the Math

Skill 37

Write a rule to explain the pattern.

1. 2500, 500, 100, 20, 4

2. 1, 15, 29, 43, 57

_____

_____

Write a rule to explain the pattern.
Use the rule to find the missing numbers.

3. 72, 63, ☐, 45, 36, 27, ☐

4. 5, 10, 20, ☐, 80, ☐

_____

_____

_____

_____

5. 9, 6, 13, 10, 17, ☐ ☐

6. 11, 14, 7, 10, 5, ☐ ☐

_____

_____

_____

_____

Check

7. Can the pattern below be explained by either the rule *multiply by 4* or the rule *add 12*? Why or why not.

    4, 16, 28, 40, 52, 64

_____

_____

_____

IIN74   Response to Intervention • Tier 3

Name _____

Patterns and Functions
Skill 38

Learn the Math

You can look for patterns to find the rule for an input/output table. You can use the rule to find missing numbers in the table.

Find a rule for the table below.

Input	5	7	8	9	10	11
Output	25	35	40	45		

Step 1 Look at the first pair of numbers.
How are 5 and 25 related?

Think: 5 + 20 = 25 So the rule could be *add 20*.

Step 2 Look at the second pair of numbers.

Test the rule *add 20* on 7 and 35.

Think: 7 + 20 ≠ 35 So the rule does not work.

Step 3 If the rule does not work, look back at the first pair of numbers.

How else could 5 and 25 be related?

Think: 5 × 5 = 25 So the rule could be *multiply by 5*.

Step 4 Test the rule *multiply by 5* on other pairs of numbers.

7 × 5 = 35   8 × 5 = 40   9 × 5 = _____

Does it work? _____

So, the rule is _____ .

_____

Use the rule to find the missing numbers in the table.

10 × 5 = _____   11 × 5 = _____

So, the missing numbers in the table are _____ and _____ .

Response to Intervention • Tier 3   IIN75

Do the Math   Skill 38

Write a rule for each function table.

1.

Input	2	14	22	30
Output	5	17	25	33

_____

2.

Input	55	45	35	25
Output	51	41	31	21

_____

Write a rule for each function table. Use the rule to find the missing numbers.

3.

Input	2	5	6	8	10
Output	4		12	16	

_____

4.

Input	100	80	50	30	10
Output	10	8	5		

_____

5.

Input	64	40	32	16	8
Output	16		8	4	

_____

6.

Input	1	30	60	89	101	135
Output	6	35		94	106	

_____

Check

7. The rule for a function table is *add 6*. Could a pair of numbers for this table be input = 3 and output = 18? Why or why not?

_____
_____
_____

IIN76 Response to Intervention • Tier 3

Name_____

Geometric Patterns

Skill 39

Learn the Math

Some geometric patterns have a pattern unit that repeats over and over.

Find a possible rule for each pattern. Then draw the next figure in your pattern.

Vocabulary

pattern unit

Notice how the figure rotates.

Rule: rotate the figure _____°, then repeat.

Some patterns have more than one rule. This pattern has two rules:

Size rule: large, small

Color rule: _____

_____

Some patterns do not repeat.

Draw the next figure.

Rule: Increase the number of sides by 1.

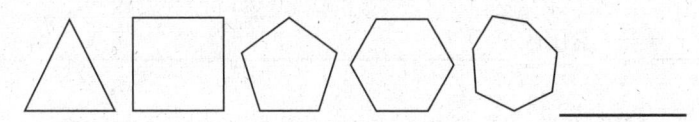

?

_____

Find a possible rule for the pattern. Then draw the missing figure in your pattern.

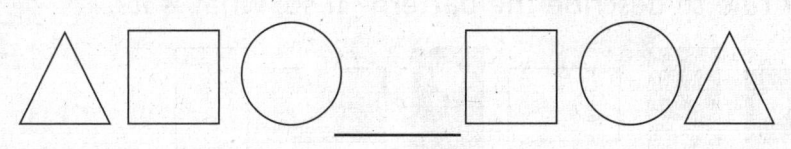

?

Rule: _____

Response to Intervention • Tier 3   IIN77

Do the Math Skill 39

Find a possible rule for each pattern. Then draw the next figure in your pattern.

1.  ▲▲△△▲▲△△ ___
   ?

   Rule: _____
   _____
   _____

2. 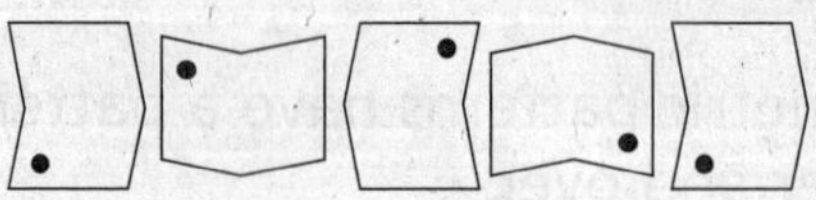 ___
   ?

   Rule: _____
   _____
   _____

3.  ☐ ☺ ☐ ☺ ☐ ☺ ___
   ?

   Rule: _____
   _____

4. 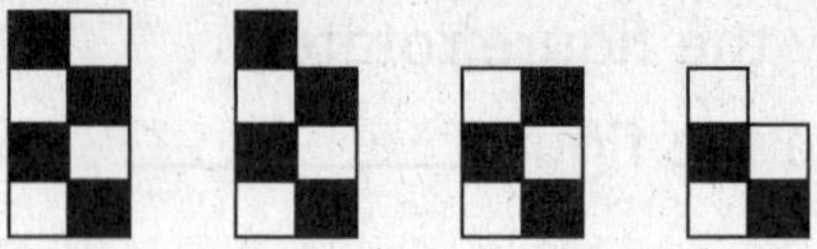 ___
   ?

   Rule: _____
   _____

Find a possible rule for the pattern. Then draw the missing figure in your pattern.

5. 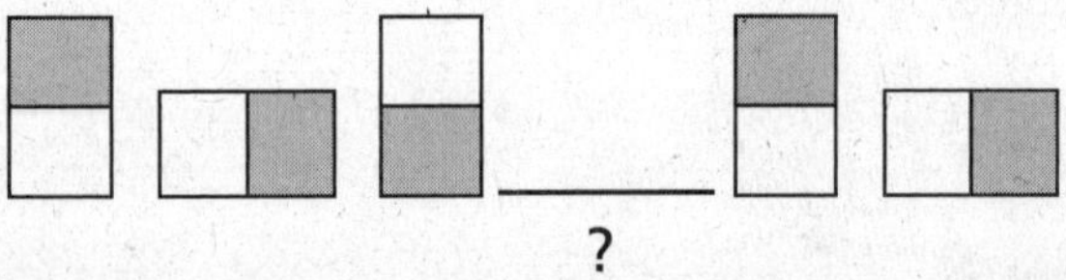 ___
   ?

   Rule: _____
   _____
   _____

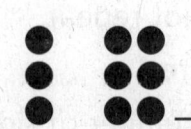

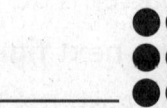

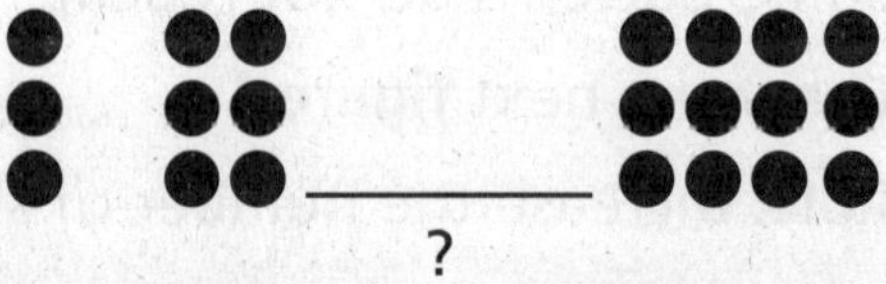

6. ___
   ?

   Rule: _____
   _____
   _____

Check

7. Aaron looked at the pattern below and said that it followed one rule: gray, white, black. Is there another rule to describe the pattern? If so, what is it?

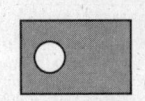

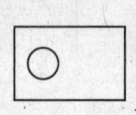

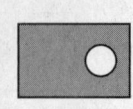

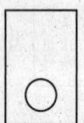

   _____
   _____

IIN78 Response to Intervention • Tier 3

Name_____

Use a Coordinate Grid
Skill 40

Learn the Math

An ordered pair is a pair of numbers that names a point on a grid. Look at the grid below. Point A is at (4,6).

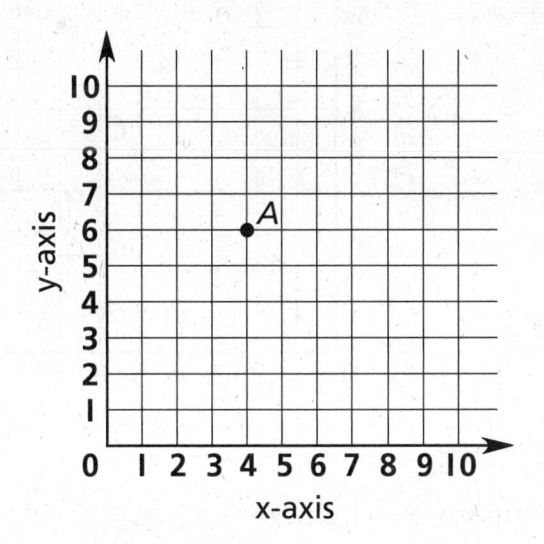

Vocabulary
ordered pair
x-axis
x-coordinate
y-axis
y-coordinate

4 is the x-coordinate. It shows how many units to move horizontally along the x-axis.

6 is the y-coordinate. It shows how many units to move vertically along the y-axis.

Use ordered pairs to locate and graph points.

Where is point A located?

Step 1 Start at (0,0). Count 2 units to the right.

Step 2 Then count 3 units up.

Point A is at (2,3).

Graph point B. It is located at (6,5).

Step 1 Start at (0,0). Count 6 units to the right.

Step 2 Count 5 units up.
Graph the point and label it B.

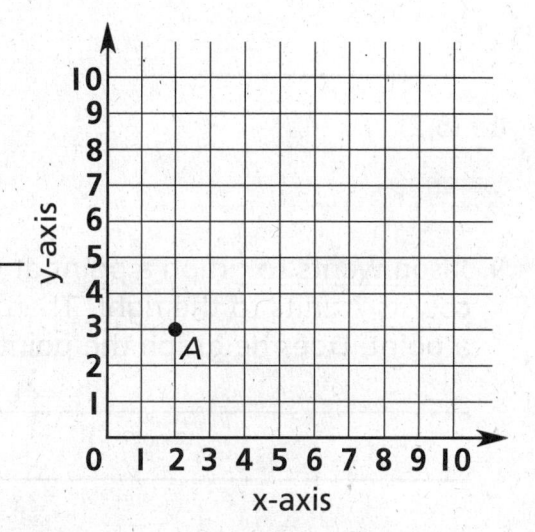

Response to Intervention • Tier 3   IIN79

Do the Math

Skill 40

Write the point for each ordered pair on the coordinate grid below.

1. (6,2)

   point _____

2. (5,5)

   point _____

3. (7,8)

   point _____

4. (0,9)

   point _____

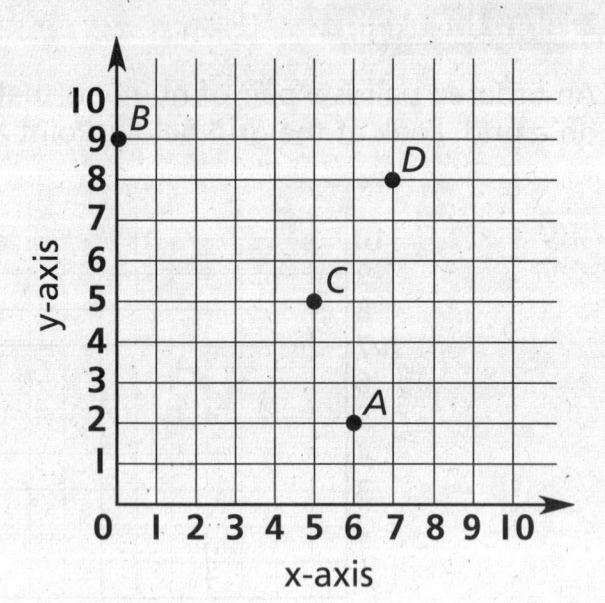

Graph and label each point on the coordinate grid below.

5. (1,4)

6. (10,3)

7. (7,7)

8. (5,2)

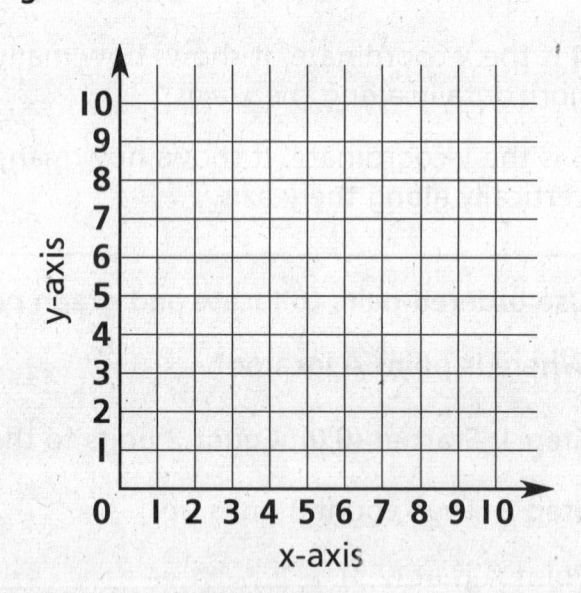

Check

9. Jason wants to graph a point at (9,7). First, he starts at (0,0) and counts 7 units to the right. Then he counts 9 units up, and graphs a point. Does he graph the point correctly? Why or why not?

_____

_____

_____

IIN80 Response to Intervention • Tier 3

Name_____

Points, Lines, and Rays
Skill 41

Learn the Math

Geometric terms are used to describe figures.

Point A point names an exact location in space.

•
A

Read: point A
Write: point A

Vocabulary
- endpoint
- line
- line segment
- plane
- point
- ray

Line A line is a straight path of points that continues without end in both directions with no endpoints.

←•————•→
  B   C

Read: line BC or line CB
Write: $\overleftrightarrow{BC}$ or $\overleftrightarrow{CB}$

Line Segment A line segment is part of a line that includes two points called endpoints, and all the points between them.

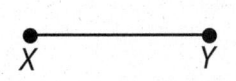

•————•
X   Y

Read: line segment ____ or line segment YX
Write: $\overline{XY}$ or ____

Ray A ray is part of a line that has one endpoint and continues without end in one direction.

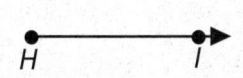

•————→
H   I

Read: ray ____
Write: $\overrightarrow{HI}$

Plane A plane is a flat surface that continues without end in all directions.

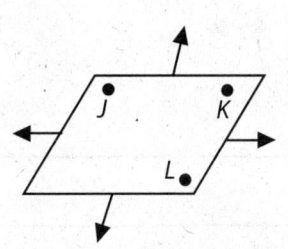

Read: plane JKL
Write: plane _____

Response to Intervention • Tier 3 IIN81

Do the Math

Skill 41

Name the number of endpoints each figures has.

1. a line

2. a ray

_____ endpoint(s)   _____ endpoint(s)

Tell whether the figure is a *line*, *line segment*, *ray*, or *plane*. Then name the figure.

3. Q—————R

4. L————→M

_____   _____

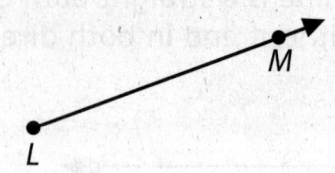

5. (plane with points T, U, V and arrows)

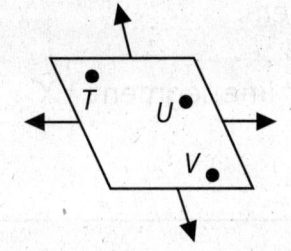

6. A————→B

_____   _____

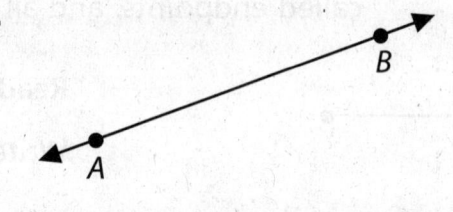

Check

7. A student made the following drawing for line segment GH. Is his drawing correct? Why or why not?

G•———•H→

_____
_____
_____

IIN82 · Response to Intervention • Tier 3

Name_____

Angles
Skill 42

Learn the Math

An angle is a figure formed by two line segments or rays that share the same endpoint. The shared endpoint is the vertex.

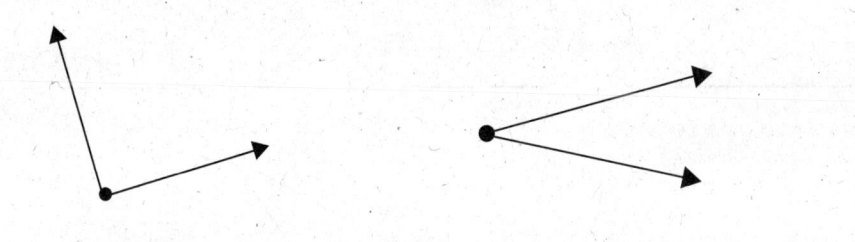

Vocabulary
- angle
- line segment
- point
- ray
- right angle
- vertex

There are different types of angles.

A right angle is a special type of angle. It forms a square corner.	Some angles are less than right angles.	Some angles are greater than right angles.
right angle	less than a right angle	greater than a right angle

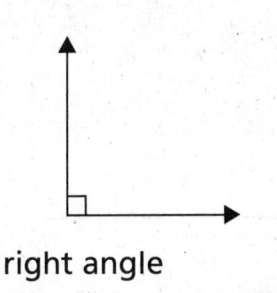

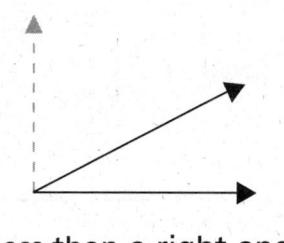

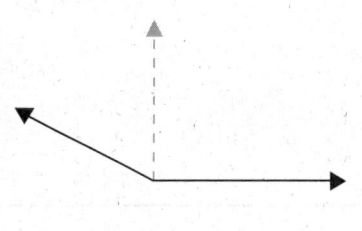

Tell whether the angle is a *right angle*, is *less than a right angle*, or is *greater than a right angle*.

Use a right angle model to compare angles. For example, a corner of a book is a right angle.

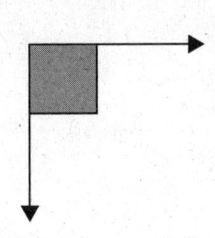

The angle is the same as the angle formed by a book corner. It is a right angle.	The angle is wider than the angle formed by a book corner. It is _____ a right angle.	The angle is smaller than the angle formed by a book corner. It is _____ a right angle.

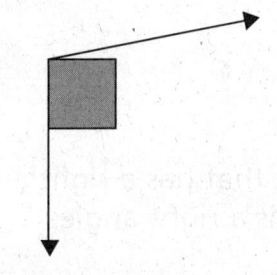

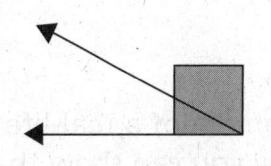

Response to Intervention • Tier 3   IIN83

Do the Math

Skill 42

Write whether each angle is a *right angle*, is *less than a right angle*, or is *greater than a right angle*.

1. _____

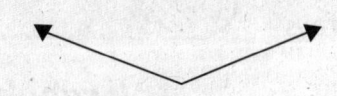

2. _____

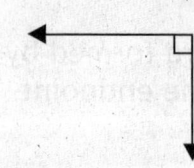

3. _____

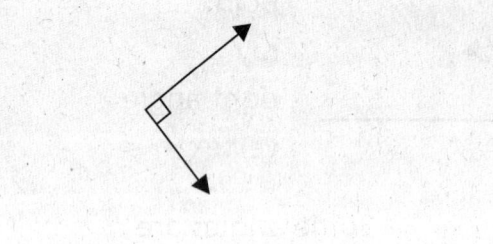

4. _____

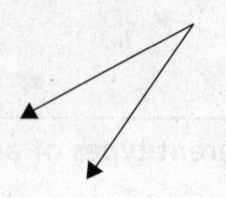

5. _____

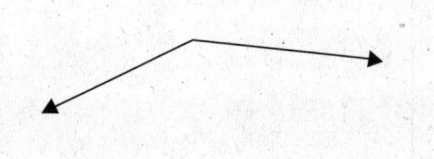

6. _____

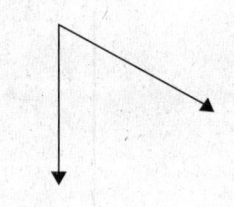

7. _____

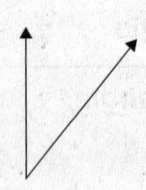

8. _____

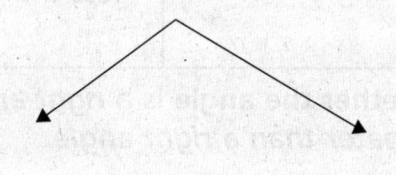

Check

9. Give an example of a real-life object that has a right angle. Explain how you can show that this is a right angle.

_____

_____

_____

IIN84 Response to Intervention • Tier 3

Name_____

Classify Angles
Skill ㊸

Learn the Math

Angles can be classified as right, acute, or obtuse.

Right Angle

A right angle measures 90°. The square marker shows a right angle forms a square corner

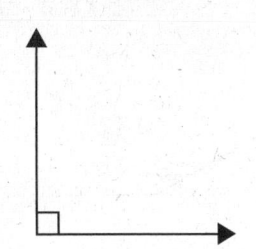

Vocabulary

acute angle
obtuse angle
right angle

Acute Angle

An acute angle is an angle that measures less than 90°.

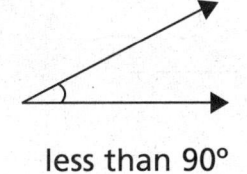

less than 90°

Obtuse Angle

An obtuse angle is an angle that measures greater than 90° and less than 180°.

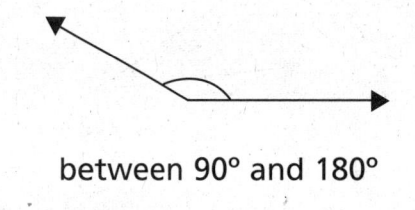

between 90° and 180°

To determine if an angle is acute or obtuse, you can compare it to a right angle.

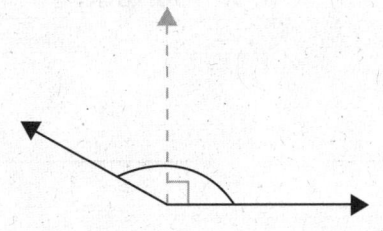

The measure of the angle is greater than 90°. It is an _____.

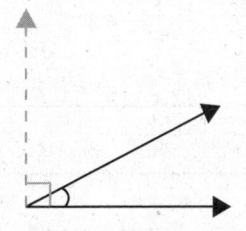

The measure of the angle is less than 90°. It is an _____.

Response to Intervention • Tier 3   IIN85

Do the Math

Skill 43

Tell whether the angle is a *right*, *obtuse*, or *acute*.

1.

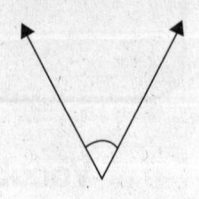

_____

2.

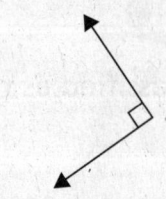

_____

3. 90°

_____

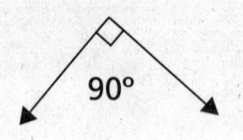

4. 152°

_____

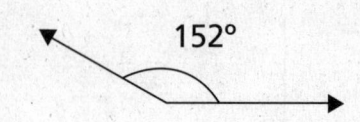

5.

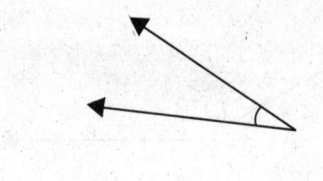

_____

6.

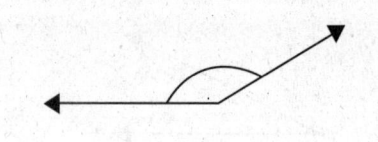

_____

7. 100° angle

_____

8. 79° angle

_____

9. 90° angle

_____

Check

10. A girl draws an angle that measures 48°. What type of angle is this? Explain how you know.

_____

_____

IIN86   Response to Intervention • Tier 3

Name_____

Polygons
Skill 44

Learn the Math

A polygon is a closed figure formed by three or more line segments.

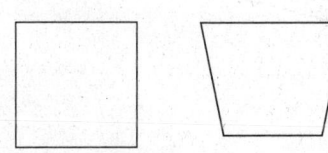

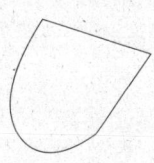

polygons   not polygons

Vocabulary

angle
decagon
hexagon
pentagon
octagon
polygon
regular polygon
quadrilateral
triangle

In a regular polygon, all sides have equal length and all angles have equal measure. A square is a regular polygon.

A polygon that has sides of different lengths and angles of different measures is not a regular polygon. A trapezoid is not a regular polygon.

Polygons are named by the number of sides or the number of angles they have.

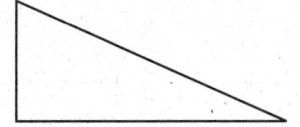

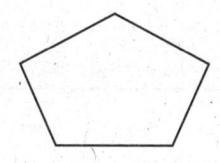

A triangle has _____ sides and 3 angles.

A quadrilateral has _____ sides and 4 angles.

A pentagon has _____ sides and 5 angles.

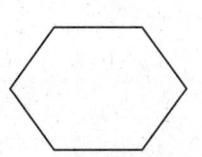

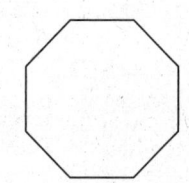

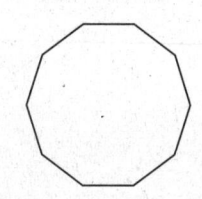

A hexagon has _____ sides and 6 angles.

An octagon has _____ sides and 8 angles.

A decagon has _____ sides and 10 angles.

What is the name of this figure?

Think: How many sides are there? _____

How many angles are there? _____

So, it is a _____.

Response to Intervention • Tier 3   IIN87

Do the Math

Skill 44

Name the polygon. Tell whether it appears to be *regular* or *not regular*.

1.

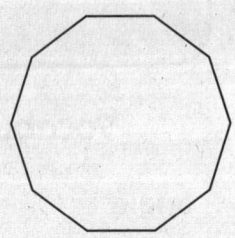

2.

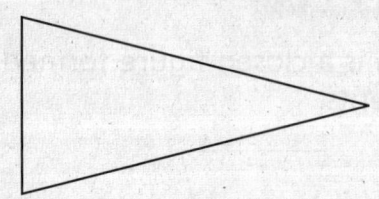

_____   _____

3.

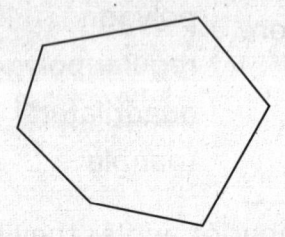

4.

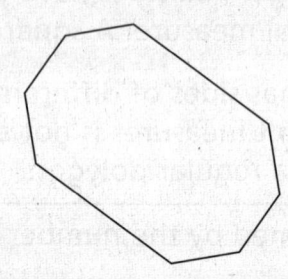

_____   _____

5.

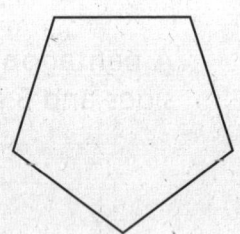

6.

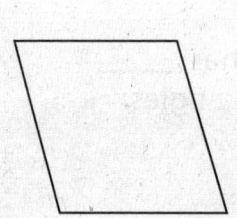

_____   _____

Check

7. Ava measured all the sides of a polygon. The lengths of the sides were 10 in., 10 in., 6 in., 6 in., and 17 in. Ava said the figure is a hexagon and is not regular. Is she correct? Why or why not?

_____
_____
_____

Name_____

Triangles
Skill 45

Learn the Math

Triangles can be classified by the lengths of their sides, or by the measures of their angles.

Classify triangles by the lengths of their sides.

	1.8 cm	2.1 cm
2 cm / 2 cm 2 cm	2.7 cm \ 2.7 cm	2.5 cm 3.8 cm
An equilateral triangle has 3 equal sides.	An isosceles triangle has exactly 2 equal sides.	A scalene triangle has no equal sides.

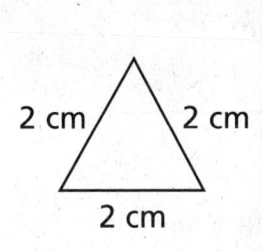

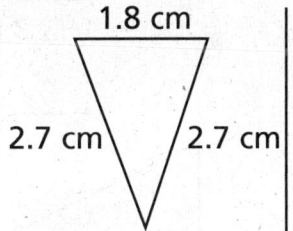

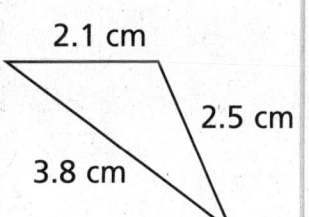

Vocabulary
acute triangle
equilateral triangle
isosceles triangle
obtuse triangle
right triangle
scalene triangle

How many equal sides does this triangle have? _____

What type of triangle is it? _____

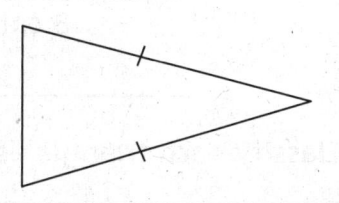

Classify triangles by the measures of their angles.

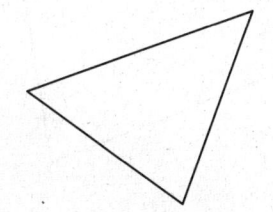

An acute triangle has 3 acute angles.	An obtuse triangle has 1 obtuse angle.	A right triangle has 1 right angle.

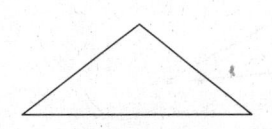

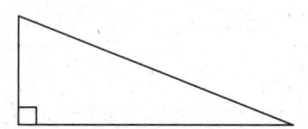

What type of angles does this triangle have? _____

What type of triangle is it? _____

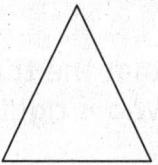

Classify this triangle by the measure of its angles and the lengths of its sides.

Does it have an obtuse angle? _____

Does it have a right angle? _____

Are any sides of equal length? _____

It is a _____ , _____ triangle.

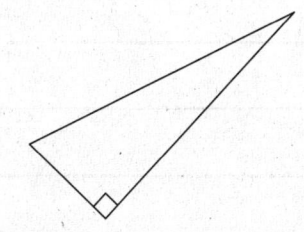

Response to Intervention • Tier 3   IIN89

Do the Math

Skill 45

Classify each triangle as *equilateral*, *isosceles*, or *scalene*.

1. 2 ft, 2 ft, 3 ft

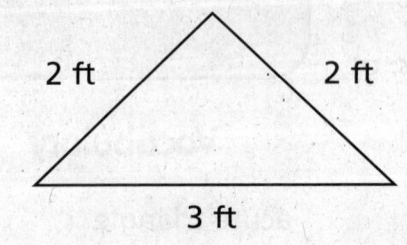

_____

2. 3 mm, 3 mm, 3 mm

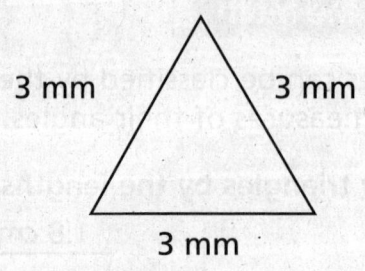

_____

3. 5 cm, 10 cm, 8 cm

_____

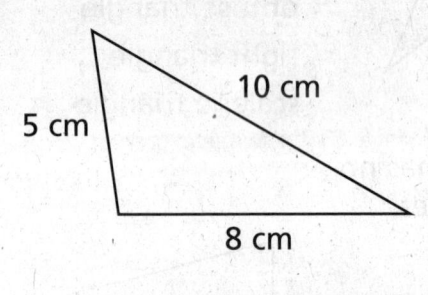

4. 2.9 cm, 6.9 cm, 6.2 cm

_____

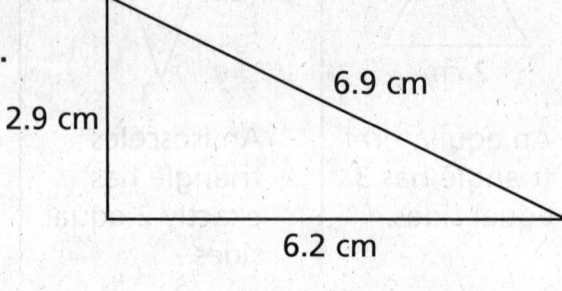

Classify each triangle as *acute*, *right*, or *obtuse*.

5. _____

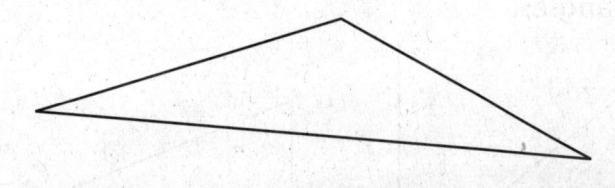

6. _____

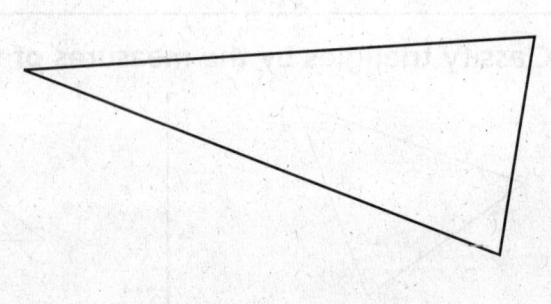

Check

7. Andrea says that the triangle below is obtuse. Robin says that it is isosceles. Who is right? Explain.

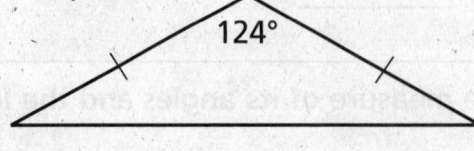

124°

_____

_____

_____

IIN90 Response to Intervention • Tier 3

Name_____

Quadrilaterals
Skill 46

Learn the Math

Quadrilaterals can be classified by the characteristics of their sides and angles.

Classify the quadrilaterals by the number of parallel sides they have.

Vocabulary
parallelogram
quadrilateral
rectangle
rhombus
square
trapezoid

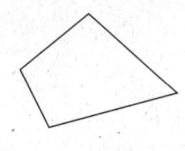

Quadrilaterals have 4 sides. This quadrilateral has no parallel sides.

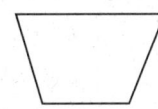

This is a trapezoid.
A trapezoid has exactly 1 pair of parallel sides.

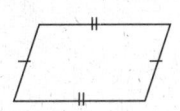

This is a parallelogram.
A parallelogram has 2 pairs of parallel sides.
Opposite sides are equal.

Classify the parallelograms.

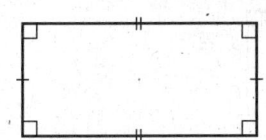

A rectangle has 2 pairs of parallel sides, opposite sides that are equal, and _____ right angles.

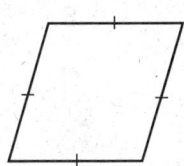

A rhombus has 2 pairs of parallel sides. It has _____ equal sides.

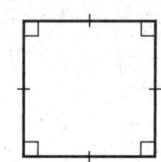

A square has 2 pairs of parallel sides. It has _____ equal sides and 4 right angles.

These figures are parallelograms because they have 2 pairs of parallel sides and the opposite sides are equal.

Is a square a rectangle?   **Think:** A square has 2 pairs of parallel sides. The opposite sides are equal. It has 4 right angles

So, a square __is__ a rectangle.

Is a square a rhombus?   **Think:** A square has 4 equal sides.

So, a square _____ a rhombus.

Response to Intervention • Tier 3   IIN91

Do the Math

Skill 46

Classify each quadrilateral in as many ways as possible. Write *parallelogram, rectangle, rhombus, square,* or *trapezoid*. If the figure is a quadrilateral only, write *quadrilateral*.

1. _____

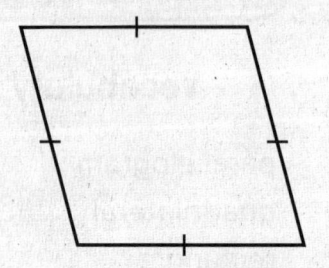

2. _____

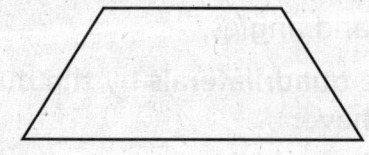

3. _____

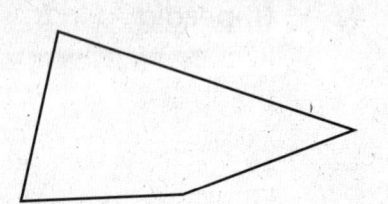

4. _____

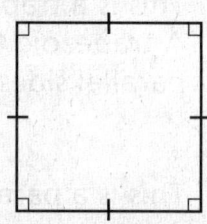

5. _____

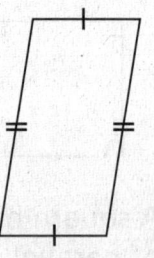

6. _____

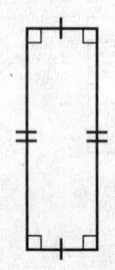

Check

7. Sam drew a quadrilateral that has 2 pairs of parallel sides. All the sides are the same length. The figure has 2 acute angles and 2 obtuse angles. Is this a square, a rectangle, or a rhombus? Explain your answer.

_____

_____

Name_____

Circles

Skill 47

Learn the Math

A circle is a closed figure made of points that are the same distance from the center. A circle can be named by its center, which is labeled with a capital letter.

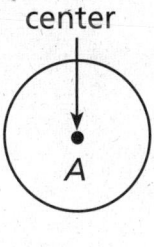

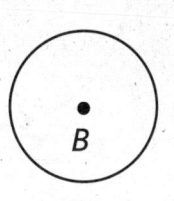

Vocabulary
- center
- chord
- circle
- diameter
- radius

circle A   circle _____

A chord is a line segment that has its endpoints on the circle.	A diameter is a chord that passes through the center of a circle.	A radius is a line segment with one endpoint at the center of the circle and the other endpoint on the circle.
$\overline{RS}$ is a chord of circle P.	$\overline{TU}$ is a diameter of circle P.	$\overline{PW}$ is a radius of circle P.

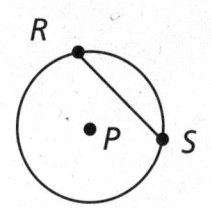

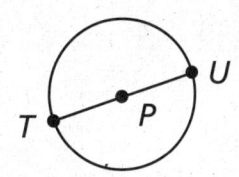

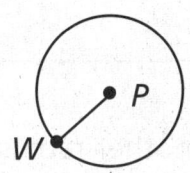

The radius of a circle is half the length of the diameter.

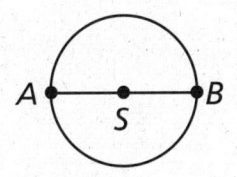

If the radius of circle S is 10 mm, what is the length of the diameter?

Think: The radius is half the length of the diameter. So, the diameter is 2 times the radius

2 × 10 = _____

So, the diameter is _____ mm.

Response to Intervention • Tier 3   IIN93

Do the Math   Skill 47

For 1–5, use the circle at the right.

1. Name the circle.

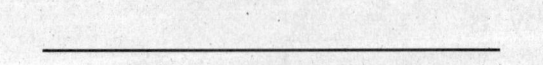

   _____

2. Name a chord that is not a diameter.

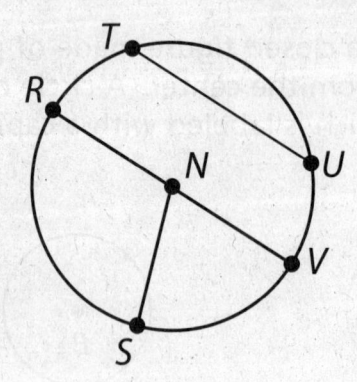

   _____

3. Name a radius.

   _____

4. Name a diameter.

   _____

5. Name another radius.

   _____

Check

6. The diameter of circle R is 4 cm. Is the radius longer or shorter than 4 cm? Explain. Then find the radius.

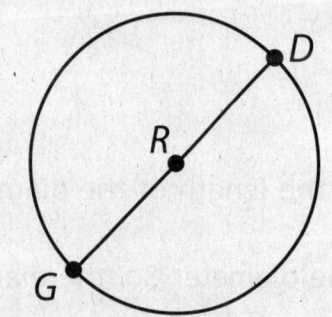

   _____

   _____

IIN94   Response to Intervention • Tier 3

Name_____

Congruent and Similar Figures
Skill 48

Learn the Math

Congruent figures have the same size and the same shape.

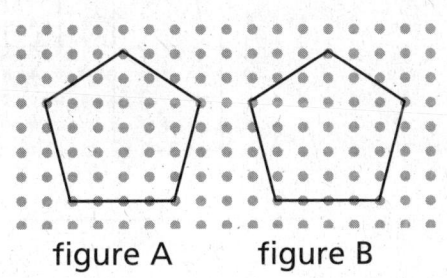

Vocabulary
congruent
similar

figure A   figure B

If you copy figure A exactly, you get figure B.
Figures A and B are **congruent** figures.

Similar figures have the same shape, but not necessarily the same size.

figure C   figure D

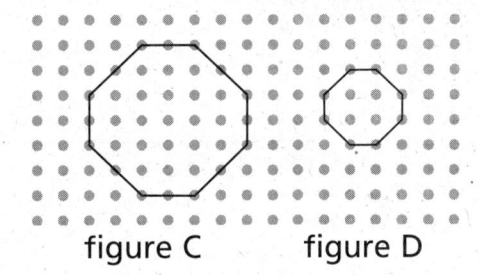

Figures C and D have the same shape. Figure D is smaller than figure C.

Figures C and D are _____ figures.

If figures are congruent they are also similar.
So figures A and B are congruent and similar.

Tell whether the figures appear to be *congruent* or *not congruent*.

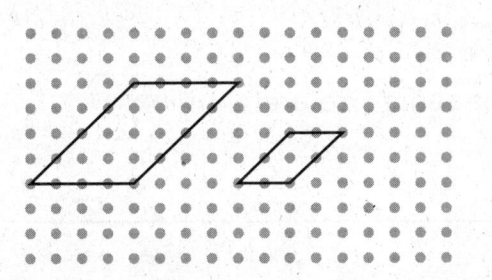

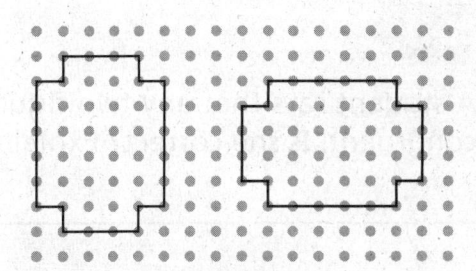

Do they have the same shape? **yes**

Do they have the same size? **no**

They are _____.

Do they have the same shape? _____

Do they have the same size? _____

They are _____.

Response to Intervention • Tier 3   IIN95

Do the Math

Skill 48

Tell whether the two figures are *congruent and similar*, *similar*, or *neither*.

1.

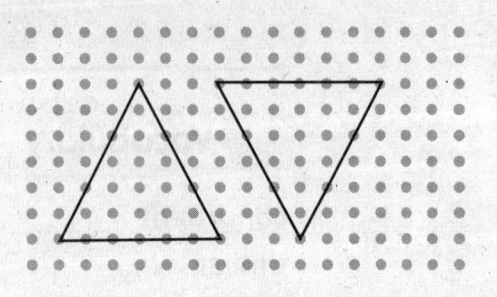

2.

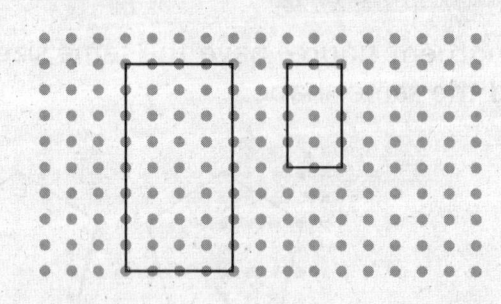

3.

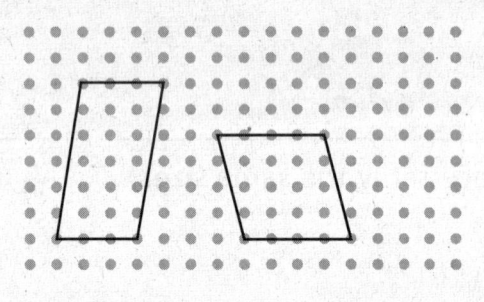

4.

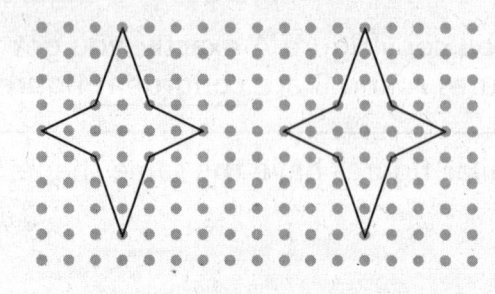

5.

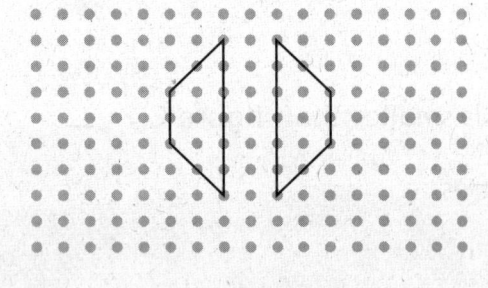

6.

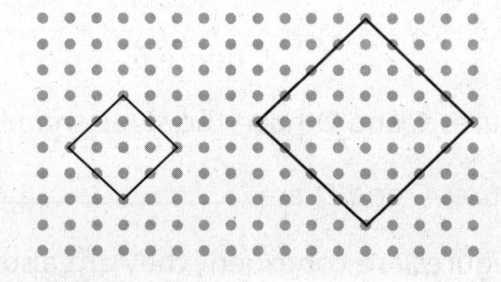

Check

7. A student says that any two figures that are the same size are congruent. Is she correct? Explain.

_____

_____

IIN96   Response to Intervention • Tier 3

Name_____

Symmetry
Skill 49

Learn the Math

A figure has line symmetry if it can be folded along a line so that the two parts match exactly.

These figures have line symmetry.

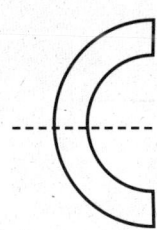

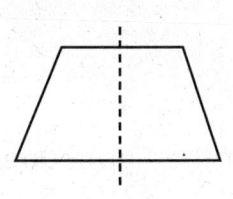

Vocabulary

line symmetry

This figure does not have line symmetry.

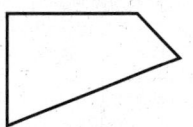

Some figures have only one line of symmetry. Others have more than one.

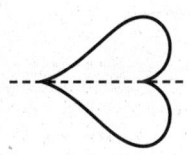

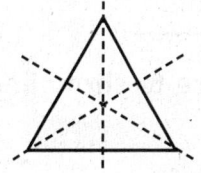

1 line of symmetry

_____ lines of symmetry

Draw all lines of symmetry for each figure.

Think: Where can the figure be folded so that the 2 halves line up exactly?

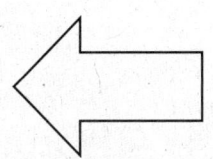

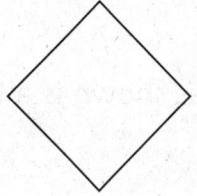

Complete each figure to show line symmetry.

Think: Imagine the figure is folded. Draw what it would look like unfolded.

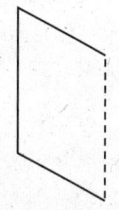

Response to Intervention • Tier 3   IIN97

Do the Math

Skill 49

Determine if the dotted line is a line of symmetry. Write *yes* or *no*.

1.

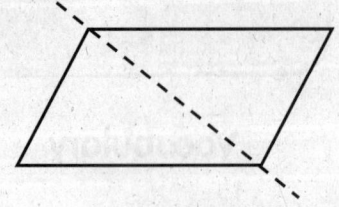

2.

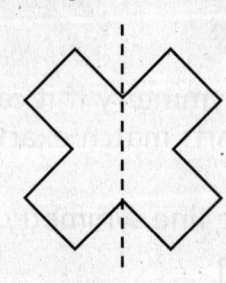

_____

_____

Draw all lines of symmetry for each figure. If the figure does not have line symmetry, write *none*.

3.

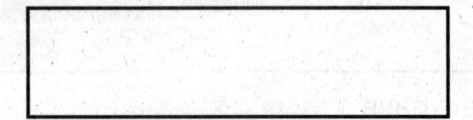

4.

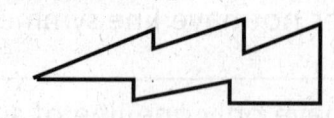

_____

_____

Complete each figure to show line symmetry.

5.

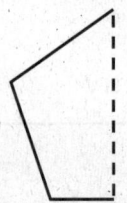

6.

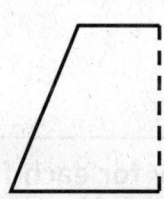

Check

7. Which of the lines shown is a line of symmetry, A or B? Explain.

_____
_____
_____

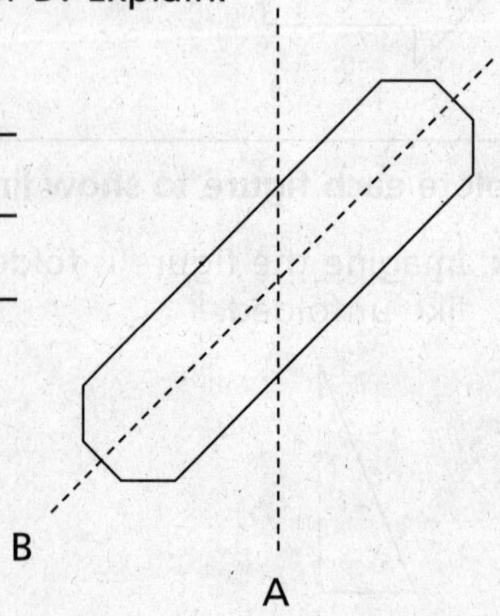

IIN98 Response to Intervention • Tier 3

Name_____

Transformations

Skill 50

Learn the Math

A transformation is the movement of a figure to a new position by a translation, rotation, or reflection.

Translation

A translation, or slide, moves a figure along a straight line to a new position.

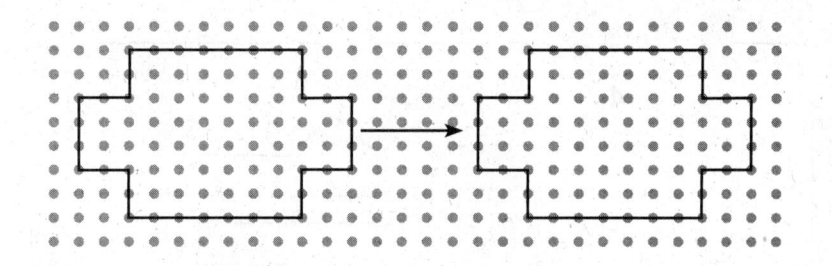

Vocabulary

reflection
rotation
transformation
translation

Rotation

A rotation, or turn, moves a figure by turning it around a point.

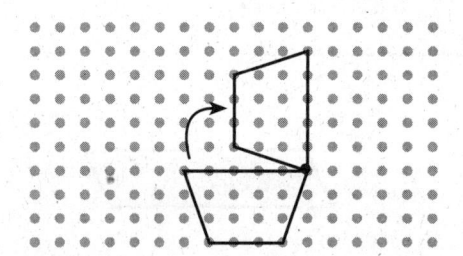

The figure is rotated 90°, or $\frac{1}{4}$ turn, clockwise.

Reflection

A reflection, or flip, moves a figure by flipping it over a line.

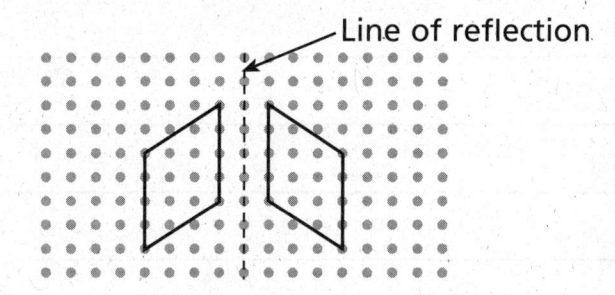

Line of reflection

Do the Math

Skill 50

Tell how each figure was moved. Write *translation*, *rotation*, or *reflection*.

1. _____

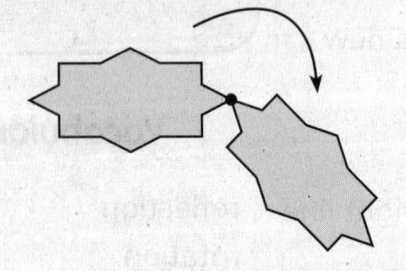

2. _____

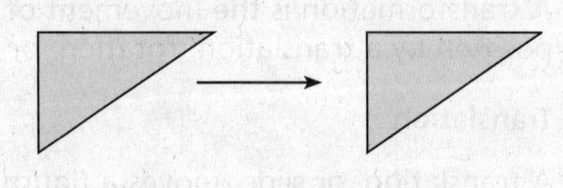

3. _____

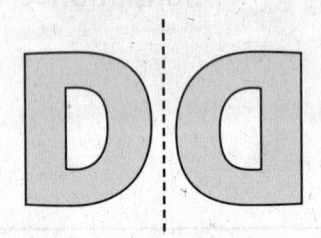

4. _____

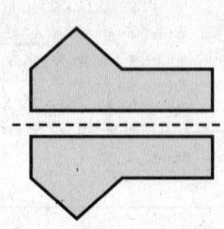

5. _____

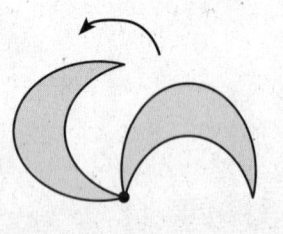

6. _____

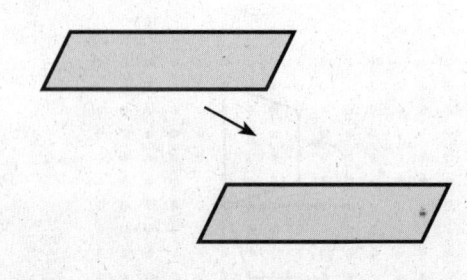

Check

7. A student draws a transformation of the original figure shown below. Is the transformation a rotation or a reflection? Explain how you know.

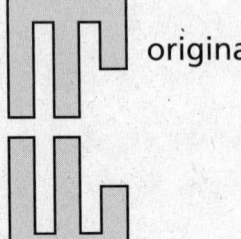

original _____
_____
_____
_____

IIN100 Response to Intervention • Tier 3

Name_____

Faces of Solid Figures
Skill 51

Learn the Math

The faces of prisms and pyramids are polygons, or plane figures.

A prism has two congruent and parallel faces called bases. All other faces of a prism are rectangles. Prisms are named by the shape of their two bases. This is a triangular prism.

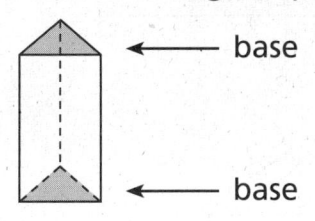

Vocabulary

base
face
prism
pyramid

← base

← base

The bases of a triangular prism are _____.

A triangular prism has _____ faces.

These are examples of other prisms.

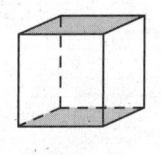

cube,
square prism

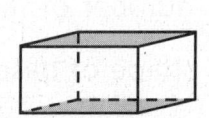

rectangular
prism

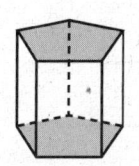

pentagonal
prism

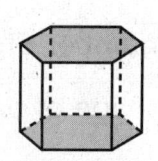

hexagonal
prism

Pyramids are also named by the shape of their bases. A pyramid only has one base. All other faces of a pyramid are triangles that meet at a single point.

This is a square pyramid.

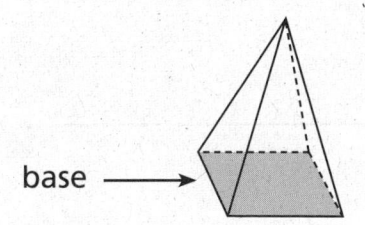

base →

The shape of its base is a _____.

A square pyramid has _____ faces.

These are examples of other pyramids.

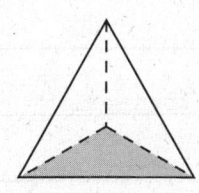

triangular
pyramid

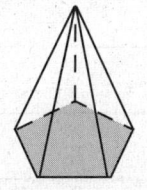

pentagonal
pyramid

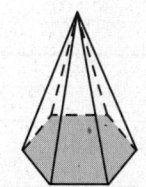

hexagonal
pyramid

Response to Intervention • Tier 3   IIN101

© Houghton Mifflin Harcourt

Do the Math　　　　　　　　　　　　　　　　　　　　　　　　　　Skill 51

Write the number of faces and the shape of the base or bases.

1. hexagonal prism

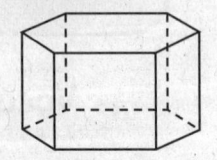

   number of faces _____

   shape of bases _____

2. rectangular prism

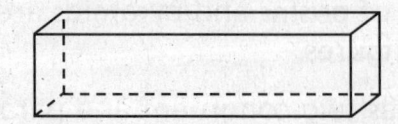

   number of faces _____

   shape of bases _____

3. square pyramid

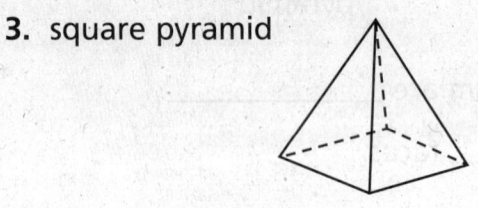

   number of faces _____

   shape of base _____

4. triangular prism

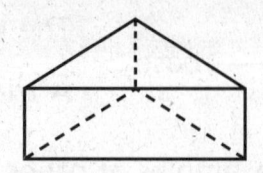

   number of faces _____

   shape of bases _____

Name each solid figure.

5. _____

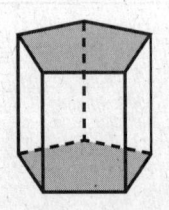

6. _____

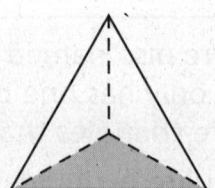

Check

7. Explain two differences between a rectangular prism and a square pyramid.

   _____

   _____

   _____

IIN102　Response to Intervention • Tier 3

Name_____

Faces, Edges, and Vertices
Skill 52

Learn the Math

Solid figures can be classified by the shape and number of their bases, faces, edges, and vertices.

Identify a face, an edge, and a vertex.

A vertex is a point where 3 or more edges meet.

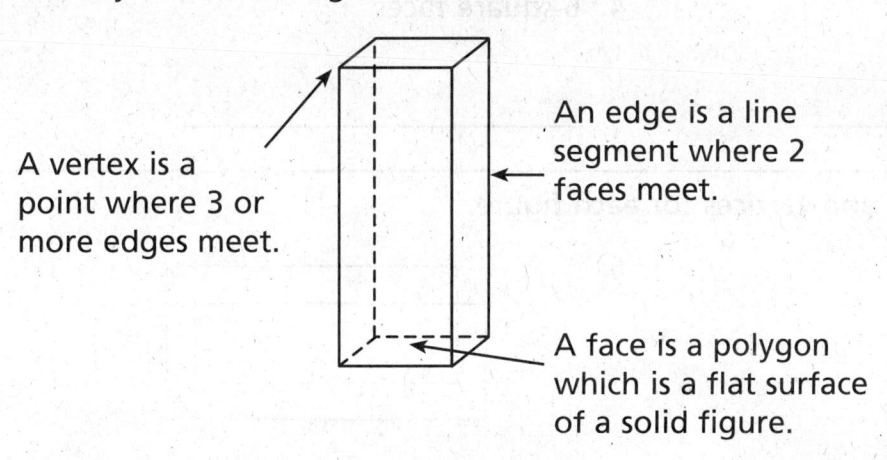

An edge is a line segment where 2 faces meet.

A face is a polygon which is a flat surface of a solid figure.

Vocabulary
- cone
- cube
- cylinder
- edge
- face
- rectangular prism
- square pyramid
- sphere
- vertex

Prisms and pyramids have faces, edges, and vertices.

cube

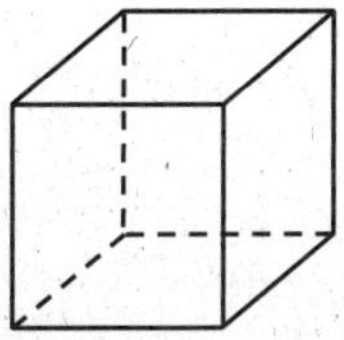

rectangular prism

square pyramid

How many faces does a cube have? ____ faces

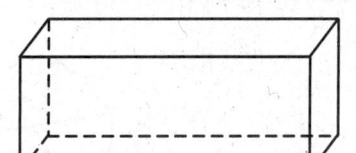

How many edges does a rectangular prism have? ____ edges

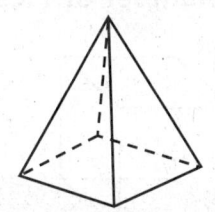

How many vertices does a square pyramid have? ____ vertices

These figures have curved surfaces and bases.

cylinder

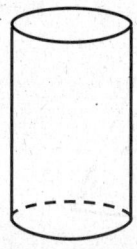

cone

sphere

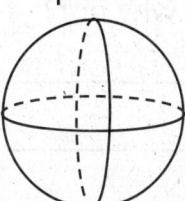

Which figure does a globe look like? _____

Response to Intervention • Tier 3   IIN103

Do the Math Skill 52

Name a solid figure for each description.

1. 8 vertices, 12 edges, 6 faces

   _____

2. 4 triangular faces, 1 square face

   _____

3. 2 circular bases

   _____

4. 6 square faces

   _____

Find the number of edges and vertices for each figure.

5.

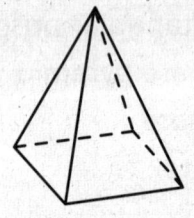

   _____ edges   _____ vertices

6.

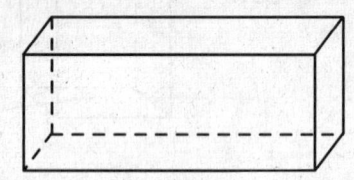

   _____ edges   _____ vertices

Find the number of faces for each solid figure.

7.

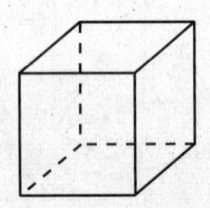

   _____ faces

8.

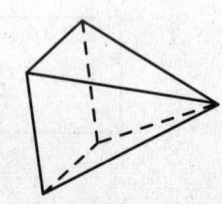

   _____ faces

Check

9. Anderson said that this figure is a cone. Is he correct? Why or why not?

   _____
   _____
   _____

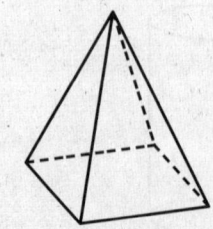

IIN104   Response to Intervention • Tier 3

Name_____

Choose the Appropriate Unit
Skill 53

Learn the Math

Length and distance can be measured in customary units or metric units.

Customary Units

Vocabulary

centimeter
foot
inch
kilometer
meter
mile
yard

inch (in.)	foot (ft)	yard (yd)	mile (mi)
The length of a paperclip is about 1 inch.	The length of a piece of notebook paper is about 1 foot.	A yard is about the length of a baseball bat.	A mile is about the distance you can walk in 20 minutes.

Choose the appropriate customary unit to measure each object. Write *inches*, *feet*, *yards*, or *miles*.

- the width of a mug   _____inches_____
- a child's height   _____
- the distance between two airports _____
- the length of a hallway   _____

Metric Units

centimeter (cm)	meter (m)	kilometer (km)
The length of a fingernail is about 1 cm.	A meter is about the width of a door.	A kilometer is about the distance you can walk in 10 minutes.

Choose the appropriate metric unit to measure each object. Write *meters*, *kilometers*, or *centimeters*.

- the height of a tree   _____
- the length of a flower petal   _____
- the distance from New York to Los Angeles   _____

Response to Intervention • Tier 3   IIN105

Do the Math Skill 53

Circle the more reasonable customary unit to measure each object.

1. the length of a banana

   inches   or   feet

2. the distance a train travels in 1 hour

   yards   or   miles

3. the length of a football field

   inches   or   yards

4. the height of a stool

   feet   or   yards

Circle the more reasonable metric unit to measure each object.

5. the length of a necklace

   centimeters   or   meters

6. the length of a highway

   meters   or   kilometers

7. the height of a book

   centimeters   or   meters

8. the height of a house

   meters   or   kilometers

Check

9. Why would it be difficult to measure the height of a skyscraper using inches? Which unit of measure would be more appropriate?

   _____
   _____
   _____

IIN106   Response to Intervention • Tier 3

Name_____

Customary Units of Capacity
Skill 54

Learn the Math

Capacity is the amount a container can hold when filled. Customary units for measuring capacity include cup, pint, quart, and gallon.

cup (c)	pint (pt)	quart (qt)	gallon (gal)
A coffee cup holds about 1 cup.	A bottle of water holds about 1 pint.	A medium-sized ice cream container holds about 1 quart.	A bucket of paint holds about 1 gallon.

Vocabulary
capacity
cup
pint
quart
gallon

Which is the most reasonable unit to measure capacity?

- a single serving of soup, cup or quart _____cup_____
- the amount of water in a bathtub, pint or gallon _____

Change customary units of capacity.

6 pints = ■ cups

To change larger units to smaller units, multiply:

6 × 2 = 12
↓ ↓ ↓
pints × number of = total
        cups in 1 pint   cups

1 pint (pt) = 2 cups (c)
1 quart (qt) = 2 pints
1 gallon (gal) = 4 quarts

So, 6 pints equals 12 cups.

Change customary units of capacity.

24 quarts = ■ gallons

To change smaller units to larger units, divide:

24 ÷ 4 = 6
↓ ↓ ↓
quarts ÷ number of quarts = total
         in 1 gallon   gallons

So, 24 quarts equals _____ gallons.

Response to Intervention • Tier 3   IIN107

Do the Math

Skill 54

Circle the more reasonable unit of measure.

1. a serving of juice

   cups   or   quarts

2. the water in a wading pool

   pints   or   gallons

3. the amount of punch a punch bowl can hold.

   cups   or   quarts

4. the amount of water a flower vase can hold.

   pints   or   gallons

Change the units.

5. 8 pints = _____ cups

6. 12 pints = _____ quarts

7. 16 quarts = _____ gallons

8. 3 gallons = _____ quarts

9. 16 pints = _____ cups

10. 20 quarts = _____ pints

11. 16 cups = _____ pints

12. 10 quarts = _____ pints

Check

13. Gina was asked to change 12 pints into quarts. She said that 12 pints is 24 quarts. Explain her error.

_____

_____

_____

IIN108   Response to Intervention • Tier 3

Name_____

Metric Units of Capacity
Skill 55

Learn the Math

Milliliters and liters are metric units of capacity.

milliliter (mL)	liter (L)
A dropper holds about 1 milliliter. A milliliter is less than a spoonful.	A plastic sports bottle holds about 1 liter of water. One liter is 1,000 mL.

Vocabulary

capacity
milliliter
liter

Which is the more reasonable unit to measure capacity?

- a dose of medication, 10 mL or 10 L   __10 mL__

- a sink full of water, 10 ml or 10 L   _____

Change metric units of capacity.

5 liters = ▪ milliliters

To change larger units to smaller units, multiply:

5 × 1,000 = 5,000
↓ ↓ ↓
liters × number of = total
        mL in 1 liter   milliliters

Think: There are 1,000 milliliters in 1 liter. Multiply by 1,000.

So, 5 liters equals _____ milliliters.

Change metric units of capacity.

2,000 milliliters = ▪ liters

To change smaller units to larger units, divide:

2,000 ÷ 1,000 = _____
↓ ↓ ↓
liters ÷ number of = total
        mL in 1 liter   milliliters

Think: There are 1,000 milliliters in 1 liter. Divide by 1,000.

So, 2,000 milliliters equals _____ liters.

Response to Intervention • Tier 3   IIN109

Do the Math　　　　　　　　　　　　　　　　　　　　　　　　Skill 55

Circle the more reasonable unit of measure.

1. a full tank of gasoline for a car

    50 liters   or   50 milliliters

2. the ink in a pen

    1 milliliter   or   1 liter

3. a spoonful of soup

    2 milliliters   or   2 liters

4. the water in a pool

    5,000 milliliters   or   5,000 liters

Change the units.

5. 7 liters = _____ milliliters

6. 4,000 milliliters = _____ liters

7. 12 liters = _____ milliliters

8. 1,000 milliliters = _____ liters

9. 26,000 milliliters = _____ liters

10. 200,000 milliliters = _____ liters

11. 16 liters = _____ milliliters

12. 13,000 milliliters = _____ liters

Check

13. Explain how you can find the number of liters in 8,000 milliliters.

_____

_____

_____

_____

IIN110   Response to Intervention • Tier 3

Name_____

Read a Thermometer

Skill 56

Learn the Math

Thermometers measure temperature using Fahrenheit or Celsius scales.

Vocabulary

degree Celsius (°C)
degree Fahrenheit (°F)

Degrees Fahrenheit (°F)

Degrees Fahrenheit (°F) are customary units for measuring temperature.

Water freezes at 32°F and boils at 212°F.

The reading on this thermometer is two marks above 65°.

The temperature shown on this thermometer is _____°F.

Read 67°F as "sixty-seven degrees Fahrenheit."

Degrees Celsius (°C)

Degrees Celsius (°C) are metric units for measuring temperature.

Water freezes at 0°C and boils at 100°C.

Some temperatures are less than 0 degrees. These are negative temperatures. The lowest temperature marked on this thermometer is ⁻10°C.

The temperature shown on this thermometer is _____°C.

Read 29°C as "twenty-nine degrees Celsius."

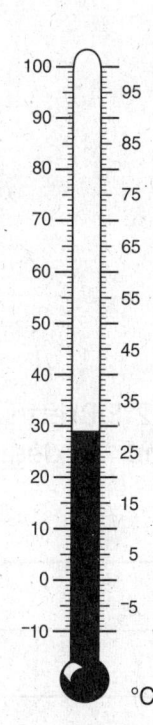

Response to Intervention • Tier 3   IIN111

Do the Math

Skill 56

Use the thermometer to find the temperature, in °F.

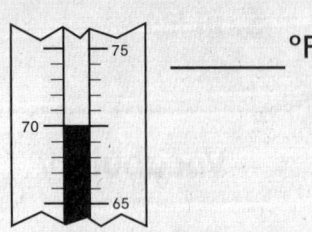

1. _____ °F

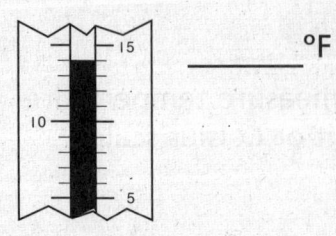

2. _____ °F

Use the thermometer to find the temperature, in °C

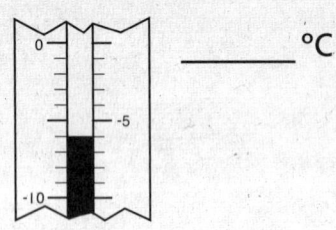

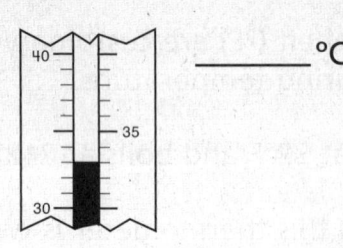

3. _____ °C

4. _____ °C

Write the temperature shown on the thermometer.
Include degrees Fahrenheit or degrees Celsius.

5. _____

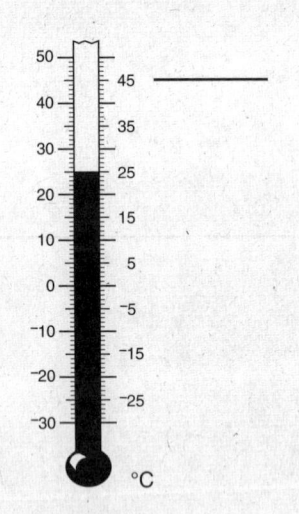

6. _____

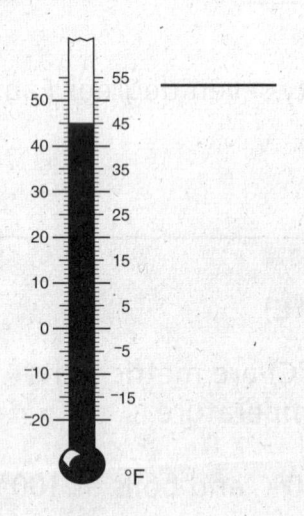

Check

7. A thermometer reads ⁻12°F. Brett says this temperature would fall between 0 and ⁻10 degrees on a thermometer. Is he correct? Explain.

_____

_____

IIN112   Response to Intervention • Tier 3

Name_____

Perimeter
Skill 57

Learn the Math

Perimeter is the distance around a figure.

You can count the number of units around the outside of the figure.

Vocabulary

perimeter

Perimeter = _____ units

You can use a centimeter ruler to measure the length of each side. Then add the lengths of the sides.

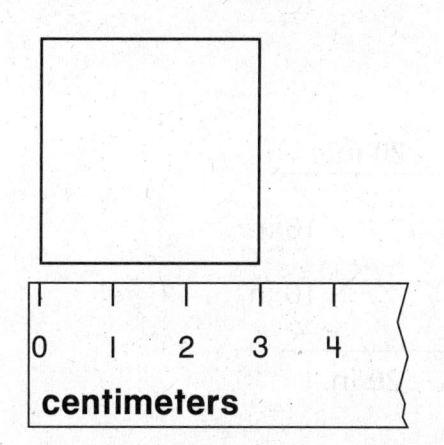

bottom ___3___ cm
right side _____ cm
top _____ cm
left side _____ cm

___ cm + ___ cm + ___ cm + ___ cm = ___ cm

Perimeter = _____ cm

You can add the lengths of the sides if you already know the length of each side.

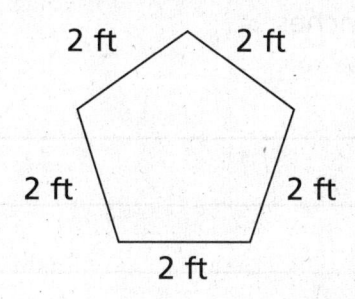

2 ft   2 ft
2 ft   2 ft
2 ft

___ + ___ + ___ + ___ + ___

Perimeter = _____ ft

Response to Intervention • Tier 3   IIN113

Do the Math　　　　　　　　　　　　　　　　　　　　　　　　**Skill 57**

Count the number of units to find the perimeter.

1. Perimeter = _____ units

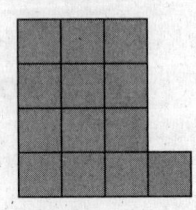

2. Perimeter = _____ units

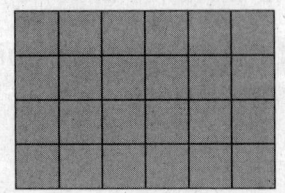

Measure with a centimeter ruler to find the perimeter.

3.

Perimeter = _____ cm

4.

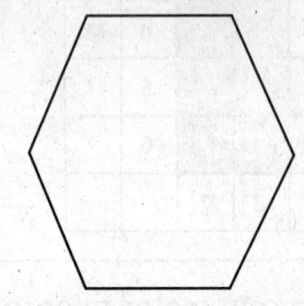

Perimeter = _____ cm

Add the side lengths to find the perimeter.

5.
- 3 in. (top)
- 4 in. (left)
- 4 in. (right)
- 5 in. (bottom)

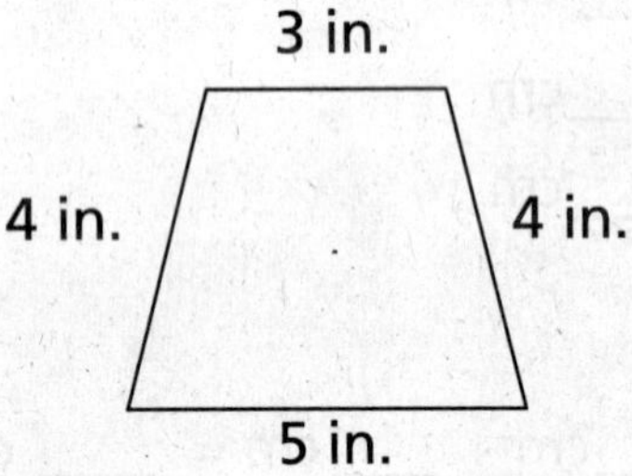

Perimeter = _____ in.

6.
- 20 in. (top)
- 16 in.
- 16 in.
- 20 in. (bottom)
- 20 in. (left)

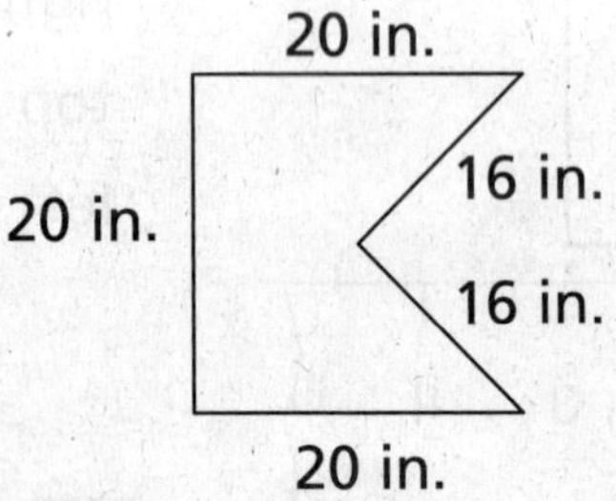

Perimeter = _____ in.

Check

7. The perimeter of a hexagon is 20 inches. The lengths of five of the sides are 5 inches, 5 inches, 3 inches, 1 inch and 2 inches. What is the length of the sixth side? How do you know?

_____

_____

_____

IIN114　Response to Intervention • Tier 3

Name_____

Estimate and Find Area
Skill 58

Learn the Math

You can find the area of a figure by counting the number of square units needed to cover the surface. You can use multiplication to find the area of rectangles and squares.

Vocabulary
area
square unit

Estimate the area of this figure.

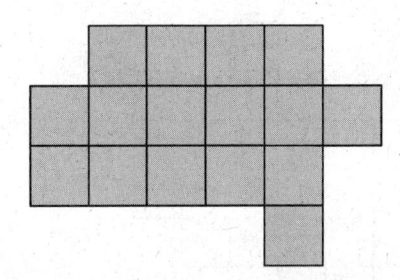

Estimate = _____ square units

Find the area of the figure by counting square units.

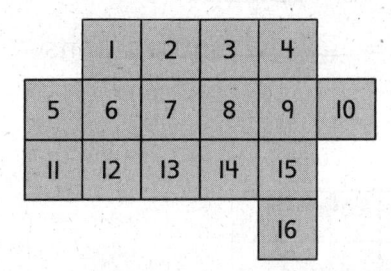

Area = _____ square units

Find the area by multiplying.

You can find the area of a square or a rectangle by multiplying the number of rows by the number of square units in each row.

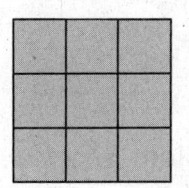

number of rows: _____

number of square units in each row: _____

Area = ___ × ___ = ___ square units

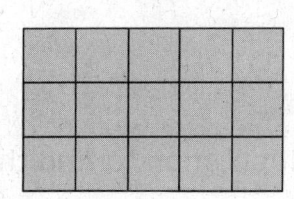

number of rows: _____

number of square units in each row: _____

Area = ___ × ___ = ___ square units

Response to Intervention • Tier 3   IIN115

Do the Math Skill 58

Estimate and find the area of each figure.

1.

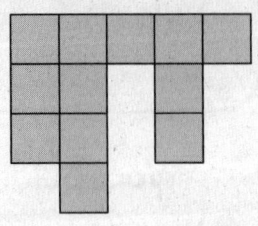

Estimate = _____ square units

Area = _____ square units

2.

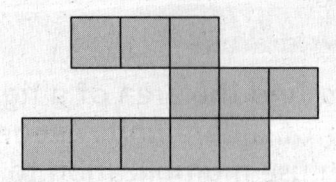

Estimate = _____ square units

Area = _____ square units

3.

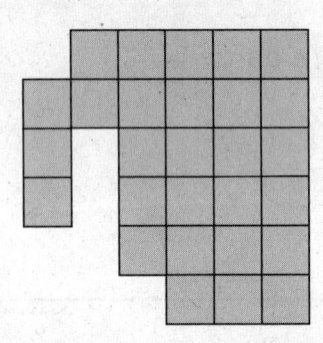

Estimate = about _____ square units

Area = _____ square units

4.

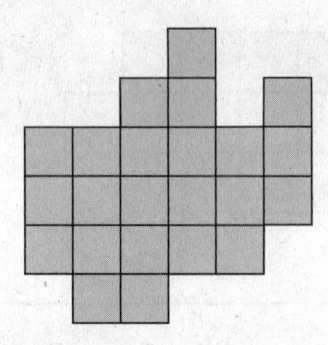

Estimate = _____ square units

Area = _____ square units

Multiply to find the area.

5.

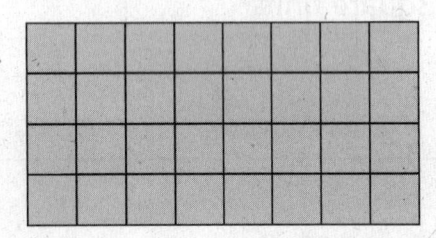

_____ × _____

Area = _____ square units

6.

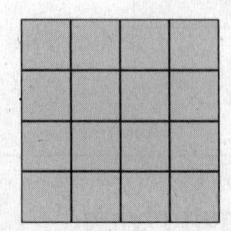

_____ × _____

Area = _____ square units

Check

7. Why can't you use multiplication to find the area of the figure in Problem 3?

_____

_____

_____

IIN116   Response to Intervention • Tier 3

Name_____

Multiply with Three Factors
Skill 59

Learn the Math

You can multiply three factors in different ways.

Multiply. 8 × 2 × 3

You can group the factors this way.   (8 × 2) × 3

Vocabulary

Associative Property of Multiplication

parentheses

Step 1 Find 8 × 2.

  8 × 2 = _____

Step 2 Multiply the product, 16, by the last factor, 3.

  16 × 3 = _____

So, 8 × 2 × 3 = _____.

___

You can group the factors another way.   8 × (2 × 3)

Step 1 Find 2 × 3.

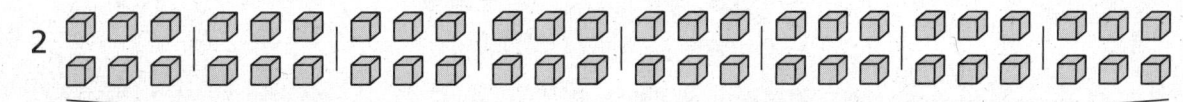

  2 × 3 = _____

Step 2 Multiply the product, 6, by the first factor, 8.

  8 × 6 = _____

So, 8 × 2 × 3 = _____.

The Associative Property of Multiplication states that you can group factors in different ways and get the same product.

So, 8 × 2 × 3 = (8 × 2) × 3 = 8 × (2 × 3) = _____.

Response to Intervention • Tier 3   IIN117

Do the Math

Skill 59

Use parentheses to show how you can group the factors. Write the multiplication sentence and solve.

1. 3 × 5 × 7 = _____
   (3 × 5) × 7
   3 × 5 = __15__
   15 × 7 = __105__

2. 6 × 1 × 4 = _____

3. 8 × 4 × 2 = _____

4. 2 × 8 × 4 = _____

5. 7 × 1 × 2 = _____

6. 4 × 5 × 5 = _____

7. 4 × 5 × 6 = _____

8. 9 × 6 × 3 = _____

Check

9. Use parentheses to show two ways to group the factors in the multiplication problem below. Then explain how to find the product for each grouping.

   3 × 3 × 8

   _____
   _____
   _____

   3 × 3 × 8 = _____

IIN118   Response to Intervention • Tier 3

Name_____

Explore Volume
Skill 60

Learn the Math

Volume is the amount of space a solid figure occupies.

Volume is measured using cubic units.

A cubic unit is a cube with length, width, and height equal to 1 unit.

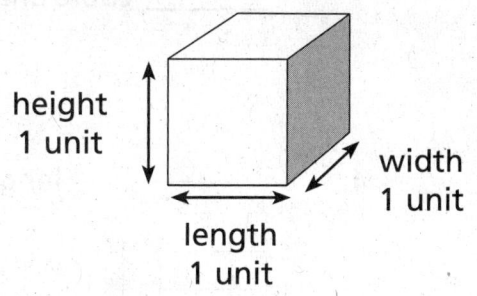

Vocabulary

volume
cubic units

height 1 unit
width 1 unit
length 1 unit

Volume is equal to the number of cubic units that can fill a solid figure.

Find the volume by counting the number of cubes in each layer. Then add.

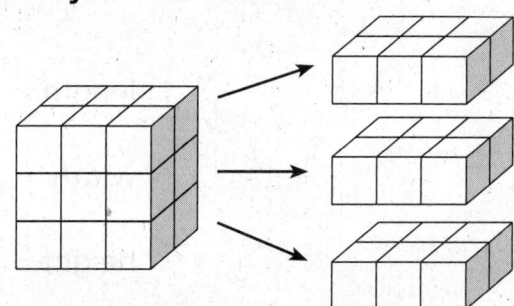

top layer = __6__ cubes

middle layer = __6__ cubes

bottom layer = __6__ cubes

Add the total number of cubes in each layer. $6 + 6 + 6 = 18$

So, the volume of the prism is _____ cubic units.

Find the volume by multiplying.

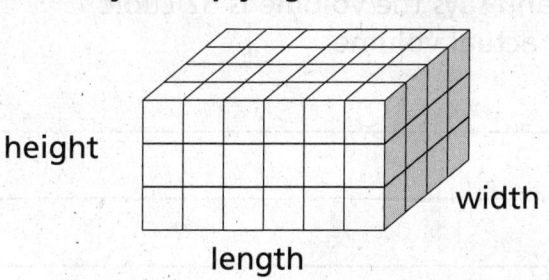

length = ____ cubes

width = ____ cubes

height = ____ cubes

Multiply length × width × height. $6 \times 4 \times 3 =$ _____

The volume of the prism is _____ cubic units.

Response to Intervention • Tier 3   IIN119

Do the Math

Skill 60

Count to find the volume.

1. _____ cubic units

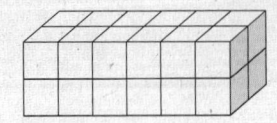

2. _____ cubic units

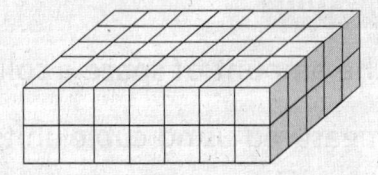

Multiply to find the volume.

3. length = ____
width = ____
height = ____
_____ cubic units

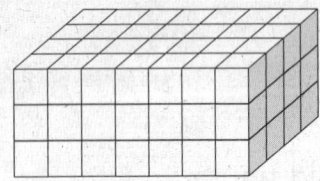

4. length = ____
width = ____
height = ____
_____ cubic units

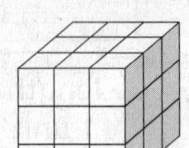

5. length = ____
width = ____
height = ____
_____ cubic units

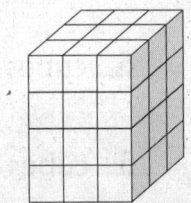

6. length = ____
width = ____
height = ____
_____ cubic units

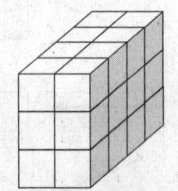

Check

7. Tim counts the cubes in this figure and says the volume is 52 cubic units. Explain his error and find the actual volume.

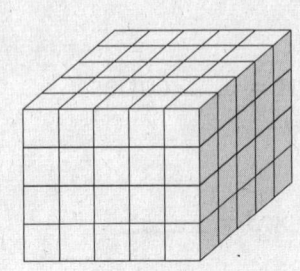

_____
_____
_____
_____

IIN120   Response to Intervention • Tier 3

Name_____

More Likely, Less Likely, Equally Likely
Skill 61

Learn the Math

An event that has more possible outcomes than another event is *more likely* to happen.

An event that has fewer possible outcomes than another event is *less likely* to happen.

Events that are *equally likely* to happen have the same number of possible outcomes

Vocabulary
more likely
less likely
equally likely

There are 2 white, 4 gray, and 2 black marbles in a bag.

Are you more likely to pull a black marble or a gray marble from the bag?

number of black marbles _____

number of gray marbles _____

There are 2 possible outcomes for black and 4 possible outcomes for gray.

Which event has more possible outcomes? _____

So, choosing gray is _____**more likely**_____ than choosing black.

Are you more likely to pull a white marble or a gray marble?

number of white marbles _____   number of gray marbles _____

There are _____ possible outcomes for white.

There are _____ possible outcomes for gray.

Which event has fewer possible outcomes? _____

So, choosing white is _____ than choosing gray.

Are you more likely to pull a black marble or a white marble?

number of white marbles _____   number of black marbles _____

There are _____ possible outcomes for white.

There are _____ possible outcomes for black.

The events have the same number of possible outcomes.

So, it is _____ that black or white will be chosen.

Response to Intervention • Tier 3   IIN121

Do the Math

Skill 61

Use this spinner to solve Problems 1–3.

1. Which event is more likely than spinning white?

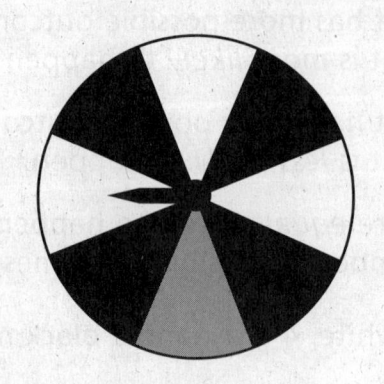

   _____

2. Which event is less likely than spinning white?

   _____

3. Is any event as equally likely as spinning white?

   _____

Tell whether Event A is *more likely* than, *less likely* than, or as *equally likely* as Event B when tossing a number cube labeled 1 to 6.

4. Event A: tossing a 2
   Event B: tossing a 6

   _____

5. Event A: tossing a 4
   Event B: tossing a number other than 4 or 6

   _____

Check

6. Explain why it is more likely to toss an even number than to toss a 5 on a number cube labeled 1 to 6.

   _____
   _____
   _____
   _____

IIN122   Response to Intervention • Tier 3

Name_____

Tree Diagrams
Skill 62

Learn the Math

Tree diagrams are organized lists that show all possible combinations. A tree diagram uses branches to connect the choices from groups of items.

Vocabulary

combination
tree diagram

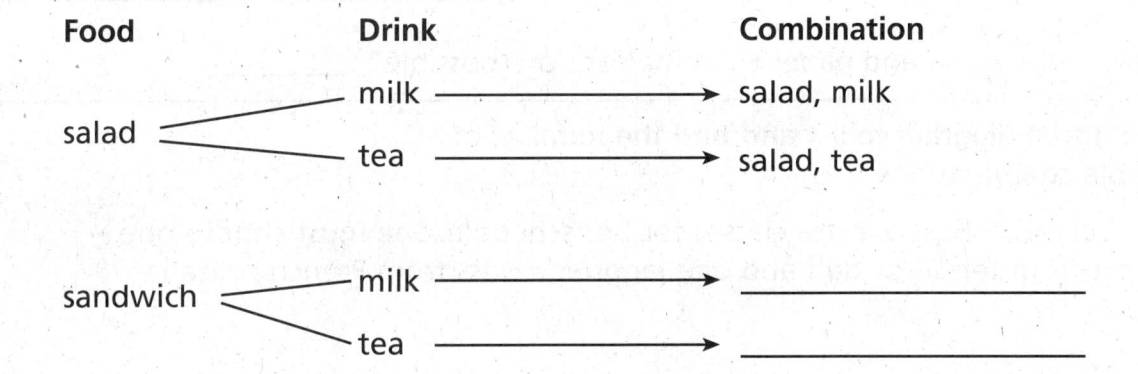

You can use a tree diagram to show all the combinations of food and drinks on this lunch menu.

Lunch Menu	
Food	Drinks
salad	milk
sandwich	tea

Food   **Drink**   **Combination**

salad ─┬─ milk ──────► salad, milk
       └─ tea ───────► salad, tea

sandwich ─┬─ milk ──────► _____
          └─ tea ───────► _____

How many combinations are there? _____

A store sells red, green, and blue pencils with round or square erasers. Complete the tree diagram and find all the pencil and eraser combinations.

Pencil   **Eraser**   **Combination**

_____ ─┬─ round ──────► red pencil, round eraser
          └─ square ─────► red pencil, _____

_____ ─┬─ _____ ──────► _____
          └─ square ─────────► _____

_____ ─┬─ round ──────────► _____, round eraser
          └─ _____ ────────► _____

How many combinations are there? _____

Response to Intervention • Tier 3   IIN123

Do the Math

Skill 62

Complete the tree diagram and then find the number of possible combinations.

1. Darryl must buy one book and one package of paper. He can choose one type of book from fiction or poetry and one type of paper from black, lined, or dot.

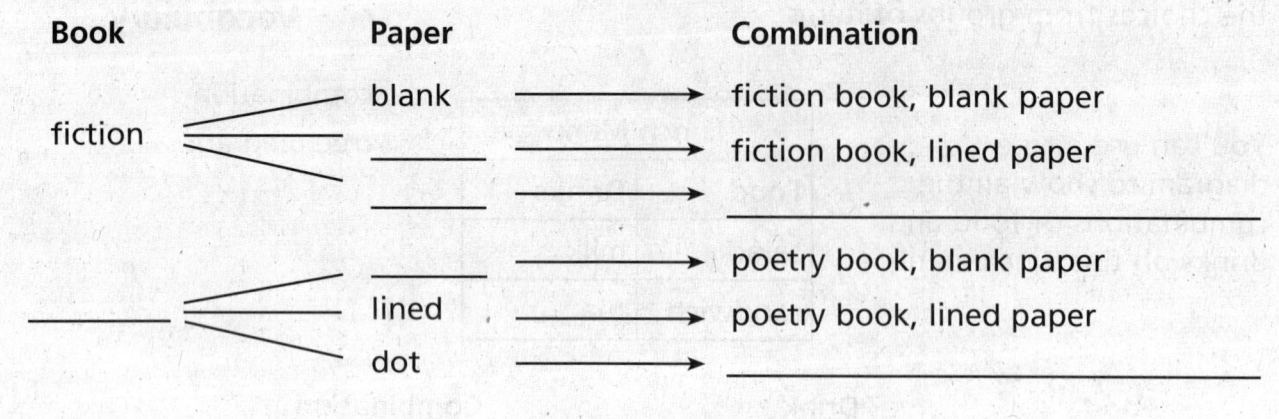

Book	**Paper**	**Combination**
fiction	blank →	fiction book, blank paper
	_____ →	fiction book, lined paper
	_____ →	_____
	_____ →	poetry book, blank paper
_____	lined →	poetry book, lined paper
	dot →	_____

How many book and paper combinations are possible? _____

Make a tree diagram to list and find the number of possible combinations.

2. Cheryl can choose 2 extra classes for her schedule. She must choose one sport from tennis or golf and one language class from French or Italian.

How many language and sport combinations are possible? _____

Check

3. Explain how you could use the tree diagram in Problem 1 to determine the number of combinations that include poetry books. Then find how many contain poetry.

_____
_____
_____

IIN124 Response to Intervention • Tier 3